SATAN SLEUTH 2

Devil, Devil

Devil, Devil

A SATAN SLEUTH Novel
by
Michael Avallone

MEWS BOOKS
LONDON AND CONNECTICUT

This one is for Joey Adams
who is always on stage
crusading against sorrow
and misery and despair

Copyright © 1975 by MICHAEL AVALLONE
First published in United States of America by Warner Paperback Library
January 1975

*

FIRST MEWS EDITION JUNE 1976

*

Mews Books are published by
Mews Books Limited, 20 Bluewater Hill, Westport, Connecticut 06880
and distributed by
New English Library Limited, Barnard's Inn, Holborn, London EC1N 2JR
Made and printed in Great Britain by Hunt Barnard Printing Ltd., Aylesbury, Bucks
45200014 9

Contents

'. . . clairvoyance, mediumship, telepathy, prophecy, possession, and all forms of parapsychological phenomena – gentlemen, I can tell you they are all one! And each and every aspect of this weird hysteria gripping the human mind and soul are all evil! In fact, the current supremacy of the Occult in this country is but a major triumph for the Devil himself . . . '

> *–Francis Parkhurst, eminent man of letters,*
> *addressing himself to the*
> *Brotherhood of ESP, April 19, 1972.*

News Item: (*Eastern Province Herald, Port Elizabeth South Africa, Dec. 11, 1973*)

MYSTERY MAN ROUTS
VOODOO TERRORISTS
'Baron Samedi'
Hoax Exposed

Haitian radicals, imported here to take a foothold among the native mineworkers with voodoo rites and beliefs, came a cropper yesterday in a bizarre finale to the recent outgrowth of terror raids and brutal murders. As hundreds of black employees gathered in the open field near Skull Head to listen to yet another re-incarnation of the legendary Baron Samedi, The Man Who Will Not Die, an unknown stranger, dressed in a priest's cassock, broke up the meeting with smoke bombs, firework displays and simple Yankee oneupmanship. The stranger challenged the Baron to a gun duel. The Baron refused as the crowd hooted its derision, trampling their Samedi voodoo dolls.

The 'priest' then engaged the 'Baron' in a stand-up, no-holds barred fight, and beat him soundly to the amazement of the onlookers. The mines will return to normal on Monday.

Voodoo is finished in South Africa and no one can say who the stranger was or where he came from.

Local authorities, on hand to regulate the wildly enthusiastic throng, report the stranger disappeared in the confusion following the fight.

More to come . . .

Book One:

SATAN'S SISTER

" . . . and the Prince of Darkness turned to the assembled women of the village of Tarnwick and proclaimed: "Cast off your worldly vestments and vow your love and allegiance to me, for you are each and all sinners and therefore you belong to me who walks in the shadows. . . ." "

– The Witch's Handbook, *pg. 9*
Salem Press, © *1793*

THE DARK ELECTRIC

Only the moon saw them.

A Manhattan moon riding high and full directly above the skyscrapered heart of the city. A moon clearly visible in a starry, frosty November sky. Yet the city was asleep. It was past five o'clock in the morning and the familiar, muted current and throbbing vitality that made New York a perpetual dynamo of released energy and unharnessed animation had now come to a momentary standstill. A time lapse which would only last until dawn and then disintegrate into the scream and rush and roar of vehicles in motion, jets in flight, construction in progress, people in pursuit of their hopes, dreams, and livelihoods.

But for now only the voice of a cold wind whispering across the stone brow of the city could be heard. Only that and the occasional mournful, lonely, and desolate hoot of a tug in the sluggish rivers embracing the narrow island of Manhattan.

The bitterly chilly early morning air did not seem to disturb the thirteen people forming a large circle by clasping their hands.

Though they were all stark naked, their contrasting bodies pale blurs under the light of the full moon, they did not belong to a nudist colony. Nor were they nudists for the sake of nudity itself.

Of course, going naked was important, essential really, if one was to be an integral part of Sister Sorrow's coven. One had to be rid of all clothing, be free of all mortal trappings and adornments, if one's flesh and skin were to become one with the Lord and Master. You cannot go to the Devil wearing the

11

vestments of the world he himself scorned and held up to scorn and ridicule. Sister Sorrow had been very clear on that point. And you had to listen to Sister Sorrow. Or you were lost, forever.

For despite all the magic and awesomeness of her appearance and earthly beauty, she was that greatest of all witches. A bride of Lucifer.

One of the Unholy Sisters. A true witch. The truest.

And here, out on the terraced patio, the penthouse garden sanctuary of her home in Manhattan, she was all things to all the people who belonged to her coven. Twenty-seven floors above the stone street, lying hidden from the ordinary mortals below, the rooftop retreat was the dark and majestic kingdom of Sister Sorrow. Lined with stunted Chinese elms, clover-leaf shrubbery, giggling pools, pebbled walks, and shrouded candle-light winking from the high walls enclosing the terrace. Candle-light encased in glass boxes. There was a heavy scent of a weird-smelling incense in the air. An aroma not unlike that of burning sulphur. A fire-and-brimstone aura.

Of darkness, decay, and death. And illusion.

The thirteen naked figures stood motionless, in a stop-freeze frame of glistening nudity, statuelike immobility. In the semi-gloom, pierced only by moonlight and the bracketed candel-abra, it was difficult to distinguish the male forms from the female. It was the utter all-alikeness, the totality of bared flesh, which rendered the grouping epicene. None of the women wore their hair long, for they were not allowed to. Only Sister Sorrow could claim full-flowing, free-falling, buttock-length hair. It wound down the sides of her triangular mask of a face in two trailing waterfalls of sheer ebony perfection, spilling along her splendidly round and full breasts before tucking away under her long arms to touch her high, arched rump. If she was a high priestess, or a witch, or a bride of the fallen angel Lucifer, she was every inch and atom the part.

That is, if an observer could ever get beyond the dynamism and striking quality of her dark, luminous eyes.

Those eyes. Twin pools of incredible luster and unknown power.

Lightning stored in two eyeballs, electric voltage imprisoned in a pair of almond-shaped windows in a divinely classic face.

The coven knew all about Sister Sorrow's eyes.

All too well.

She could transform you into a cat or a dog or a bird if she chose to. Just by looking at you and commanding the metamorphosis, if you were stupid enough for betrayal or disobedience – or *disbelief*.

She could make you a block of wood or a stone idol or a chair or a rug for people to walk on if she so desired.

If she willed it.

If she ordained the change.

Sister Sorrow was the genuine article.

The Daughter of Darkness.

And her gatherings of the coven were something to be remembered. Something to think about and wonder over all the rest of your life. If you belonged to her – were a member of the Sorrow Coven.

Like tonight. Another meeting, like all the others.

Yet tonight was different somehow. Truly unique.

The witching hour had long since passed. Twelve midnight was only a memory. The Witch's Sabbath had been fulfilled, with all the main points of ritual disposed of. The demonological *Sabbat* had been fulfilled hours ago. The Invocation, the Lord's Prayer recited backward, the sacrificial rite, the Devil's Creed – all of that Sister Sorrow had caused to be done. To be performed with true, fervent, Satanic zeal. Even the sign of the cross, rendered in reverse by the unholy thirteen, had elicted twice the usual amount of devotional joy. But –

Sister Sorrow had somehow refrained from disbanding the coven.

She had lingered on, nude and eternally spiritual and regal, invoking the forces of darkness and no one had dared interfere.

Limbs had grown weary, bodies ached with orgiastic release, eyes were straining in the half light. The aroma of incense was now overpowering but not a soul allowed hands to unlock, fingers to unclasp.

It remained for Sister Sorrow to announce the end of things.

This was her kingdom, her empire, and no mere acolyte

13

devotee could challenge her authority. Not on her home grounds. It was unthinkable.

Still – what was so different about this night?

Why had this meeting of the Inner Circle suddenly become unlike all other nights? Why the lengthy time period? What was wrong?

Or, perhaps more importantly – *what was going to happen?*

Then, at last, Sister Sorrow provided the answer.

'Who breaks the chain, who disrupts the glory of the circle, must be punished. One who accepts our hand in the Ritual of the Ring and does not truly believe – that one must die.'

Twelve startled faces, stirred from weariness and spent ecstasy, swung toward the stunning, palely beautiful face at the apex of the circle. Sister Sorrow's words, spoken low, yet somehow echoing with the latent force and power of her personality, hung in the incense-filled atmosphere.

Sister Sorrow folded her long arms across her bosom, arched her head back, and closed her eyes. It was as if a blinding illumination had winked out. The darkness of the roof patio garden was now intense. Not even the full moon could rival Sister Sorrow at her early-morning best. Dawn was but an hour away. The whispering wind seemed louder. Closer.

'There is one among us who does not believe.'

Amazed glances were exchanged. Wonder raced like the wind.

'He defies our law, our God, our Lucifer.'

Shock ran through the group but no one could speak.

'He who must die is with us now.'

Sister Sorrow's magnetic voice, deep and lilting, sent shivers down naked spines. Her tone rose suddenly, on a note of anger, but contained, for all of that. Like a litany, the condemnation edged on.

With steel and fire and ice gilding the words.

'He does not have to reveal himself. I know who he is. I, Sister Sorrow, who have been given the power to know all things.' There was a long pause in which twelve hearts lost their normal beat. *'Step away, all of you. Move off from the defiler of our sacred ceremony. And gaze at him. Gaze at the one we brought into our midst and accepted as a Brother of*

14

Darkness but now must drive from the gates of our paradise.
I will now name him, he will suffer for his great sin. . . . Do you
hear me? All of you? This is what will happen to anyone who
dares turn his allegiance from us and worship false gods. . . . Do
you hear me. . . . Brother Carmody?'

The gasp, a blurt of air in concept, a wonderment of moaning
disbelief and fear, all exploded in the heavy silence of the
penthouse roof.

All eyes spun toward the accused. The flickering candlelight
wavered.

All forms, male and female, trembled with mingled amazement and terror. Sister Sorrow's wrath was not something to
be taken lightly.

Something not to be seen – or ever forgotten.

Brother Carmody knew that too.

Perhaps better than anyone.

Abruptly, he had pushed back, moving away as if to run. But
Sister Sorrow had opened her eyes again. Suddenly, flashingly.
And two gleaming, icy beacons of vision were now trained
upon the tall but corpulent man who had but recently joined
the coven. A man about whom the others knew nothing. Save
that he was a follower, a believer, an acolyte, as they were – and
now, if Sister Sorrow was right, if what she said was true, he
was nothing . . . a doomed man who had broken the Inner
Circle.

'No . . , ' Brother Carmody moaned, his ruddy face working
in the half light. 'I – I – have done nothing. I just don't want
to do this anymore – it isn't right – it's unholy – I was out of
my mind because my wife died . . . and I was looking for
something to believe in – anything. But this ungodly stuff isn't
it. But don't worry about me – I won't tell anyone about your
little secret. God knows it's your own business.'

'Shut up, you fool,' one of the coven muttered in a low hiss
of derision. She was a slender, attractive blonde with a wide-
hipped, full-bosomed, beach-tanned body. 'You've dug your
own grave! You never should have walked into this if you
weren't going to stick with it.'

None of the others were capable of speech. Their attention
was riveted on the fantastic form of Sister Sorrow and the

15

trembling Brother Carmody, knowing, sensing, that there was magic and sorcery in the night.

In the very air itself. The incense vapors were stifling now.

'*Sister Lilith has prophesied your doom, Brother Carmody.*' Sister Sorrow raised a long white arm and leveled it toward the tall, corpulent man as he retreated, as if to exit from the circle via the French doors showing dimly behind him. '*You cannot run from my judgment, Brother Carmody. Since you have chosen the role of disbeliever and hypocrite, denying our Lord of Darkness as well as we who were your brothers and sisters, it is therefore fitting that you become that which is the lowest form of life in the animal kingdom. . . . Go down in the slime!*'

'No!' Brother Carmody bellowed, thrusting out his arms. His face was a mixture of contempt, fright, and confusion. 'Cut that stuff out, now! You've got a great act, lady, I got to admit that. But I'm walking out of here the way I came in and if you know what's good for you – all of you – you'll let me go. I tell you – '

He stopped talking, breaking off like a record suddenly stilled.

He had looked into Sister Sorrow's eyes. Not meaning to.

Even from a distance of over twenty feet he saw their meaning, read their message. He uttered a low cry, spun about, and lurched for the French doors. Sister Sorrow pointed her long arm, her incredible eyes shone, and the coven stayed on in complete awe. The tall pale blur of her superb figure was truly queenlike. No, goddesslike . . . unreal.

Something flared in the weird half world of the rooftop.

A lambent glow, a burst of fiery, smoking flame. And then the glow was gone, the flame was gone. The breath of the wind quieted.

And Brother Carmody was gone also. As if he had never *been.*

Where he had stood, no man could be seen.

The coven blinked, hearts stopping, souls turning cold, minds paralyzed with the miraculous evil of it all. The Truth of the Power.

A small, wriggling green snake now swirled across the tiles

16

before the French doors. In another second it too was gone, disappearing into the shrouded environs of the stunted Chinese elms. Like a wraith, a figment of the imagination, a nightmare. Squirming out of sight.

'*Let him go,*' Sister Sorrow murmured without inflection, '*to perhaps find the Tree of Knowledge again. Then he will know better — it is a fitting termination for a brother who should never have been welcomed to our midst. And now let us make the Sign and disband — until the Master calls for us once more — emmay, tolum, drekko — Ave Satanas!*'

As the long-awaited farewell emanated from her carmine-painted lips, the coven stared at her in newer, fresher wonderment and terror.

With her amazing eyes shut once more, Sister Sorrow was indeed everything and all that the coven believed her to be.

A witch.

A bride of Lucifer.

The Daughter of Darkness.

Twelve heads lowered.

The sign of the cross was executed, in reverse.

Twelve voices blended with Sister Sorrow's in the final litany.

The Black Mass drew to a close. The coven separated.

Not one of them disbelieved that Brother Carmody had been transformed into a green garden snake through the magic and sorcery of Sister Sorrow. Not for a single, foolish instant.

It would have been disastrous to believe otherwise.

Sister Sorrow's coven had been hand-picked, by the witch herself, for their complete and total commitment to the Prince of Darkness.

Brother Carmody was simply a mistake. An acolyte who had weakened, somehow. Who had lost the call, who had strayed from the dark and narrow. That was a mistake which must not be repeated. In the name of Satan, no!

So thought Max Toland, a banker. As did all the others.

Wayne Rilling, theatrical agent.

Norma Carlson, newspaperwoman.

Bedelia Aaron, dietician.

Sloane Gilley, Broadway producer.

Duane Farmer, news commentator.

Alice Ainsworth, magazine editor.

Jason Browne, architect.

Buck Benson, engineer.

Wanda Ravelli, school principal.

Baldwin Simpson, veteran character actor.

Sophie Smythe, commercial artist.

Dress manufacturer Charles Carmody hadn't used his head at all.

No one who was afraid of the unknown should ever go up against Sister Sorrow. Not ever, not at all. It was like committing suicide.

A genuine witch with a genuine coven in the heart of Manhattan.

Right under everybody's nose.

Everybody's.

And the beauty of it was, so many people didn't believe in witches.

The fools.

WITCH HUNT

At the Manhattan offices of the Federal Bureau of Investigation, agents Arthur Fling and Terry Lenning were knocking heads together over what had become an old argument between them. The wave of the Devil-made-me-do-it murders in their district.

Lenning was pigheadedly persistent in his theory that the whole killing spree was an organized affair whose *raison d'etre* he was unable to provide just yet. Fling didn't see it that way at all. After all, what with Watergate, the *Exorcist* cleaning up at movie box offices all over the land, and everybody and his uncle trying to find a meaning to life beyond mere scratching for a buck to live, it was no wonder to him that people were going to the Devil. Just as he had told Lenning when the series of Satan-marked deaths began – 'It's just another crackpot way of trying to find God, Terry. Stop knocking yourself out, will you? There's no mastermind behind all this. The citizens are going off the rails. No more, no less.'

Lenning wouldn't let go of an idea once he latched on to it. Fling was older, smarter, more experienced, but he knew when to give a subordinate some slack. Lenning had too much on the ball to be squashed for trying. Anyway, you could never tell about cases.

Any kind of a case.

So despite Fling's injunction to forget the Satan plot, Lenning had spent a great deal of time coordinating the files, hanging around the clacking teletype machine for reports from all over the country, and he had even gone so far as to whip up a prospectus. Charts, timetables, detail lists and all. Lenning was so eager-beaver about the whole thing that Fling's patience should have worn thin. The Manhattan desk was just too heavy with work and Lenning was too valuable to waste so much brain-power and manpower on kook theries. But Fling remained cool even though he wasn't that certain that the whole ballgame was an FBI matter in the first place. Real crime and real criminals were real enough without having to ring the Devil into the scenery. Jeezis.

'Terry,' Arthur Fling said, trying not to be unkind; it wasn't his job to stifle initiative or step on enthusiasm, 'what's got you so hopping about this cult and Devil-worship stuff anyway? Come on. Level with me, now. I won't pull rank on you.'

'Art, I'm sorry to bother you with all this but I can prove a connection in all these kills.'

'Don't be sorry.' Fling smiled from his desk. 'Just prove it.'

Lenning grinned, laying his charts and lists down on the desk top. There was no mistaking the manhunt look in his alert eyes. That look only came when an agent was onto some-thing. When he smelled the game. Fling frowned, refusing to stare at the maps and charts.

'Make your point, Terry. Stop acting like you're having kittens. It's unbecoming in a Federal agent. And I haven't got time to read just now. Say what you want to say.'

'All right, I will.' Terry Lenning folded his arms. His smoothly attractive face was now blander. 'Four deaths. One on West End, one on Riverside Drive, the third on Houston Street downtown, and the last – right in the middle of Central Park. Four women. All young. All found naked and mutilated. All killed by decapitation. All found by the police in positions of crucifixion, spread out on their backs – '

'Please. Don't remind me. I just had lunch. And we know all that. So what else is new?'

'So this: All of those females were beautiful, unmarried, not even a lover or a boyfriend the Homicide Squad could pinpoint. And the scenes of the crimes – an unrented flat on West End, a vacant apartment on the Drive, an abandoned, condemned building down there on Houston, and, of course, wide-open Central Park, just a stone's throw from Shakespeare Gardens

– a little bower of some kind, like a circle – a regular arena. You still with me?'

'To the bitter end,' Fling sighed, 'and still waiting for something I don't know. What's the punch line?'

Terry Lenning seemed to put his teeth together. 'The mutilations were superficial cuts on the bodies of all four women. But of strange design – occult marks, in fact. An ideograph expert in the Homicide Department claims the pattern of the slices all belong to the Devil-worship school of lunacy. Okay?'

Fling snorted. 'Okay, okay.'

'Okay,' Lenning affirmed, mildly enough even though he was interrupting a superior and strangely holding back the point he wanted to make. 'Now for the clincher. No witnesses to any of these kills. No clues except the corpses themselves. Not so much as a cigarette butt, a bit of torn material, or even a footprint on the floor or ground. So – my first point. Even though none of the corpses knew each other, even though no possible connection has been made between the four of them, there is one common denominator.'

'Besides the four headless bodies and the Satan markings, you mean?' Fling inserted with a note of scorn in his voice.

'Besides that. I've brushed up on my demonology and witchcraft, Arthur. And the decapitations clinch it. That and the four scenes of the crime. Four isolated spots – an unrented flat, a vacant apartment, a condemned building, and Central Park at midnight. Places where no other people were likely to be around. Our four lovely dead ladies were all *participants* and, as it turned out, *sacrifices*, in Black Mass rituals.'

Fling rocked forward in his swivel chair, scraping the floor in his anger. 'Cut it out, Terry. You've got no proof of that at all. I'll grant you the murders were close together in time – all in the same two-week period – and somebody chopped them up the same way, but don't go overboard on this Satan crap, will you? The Department doesn't care for that angle at all. Perversion and insanity will do for us very nicely, thank you. We don't need any spook occult stuff to louse these cases up.'

'Sorry, Chief,' Terry Lenning said very quietly, unruffled.

'But that's what we've got. A practicing Satan cult. The whole schmear. Because, you see – '

'*What?*' roared Arthur Fling. 'Exactly *what* do you want me to see?'

'When there's a sacrifice in the ritual of the Black Mass,' Terry Lenning said coldly, 'the custom is – no, make that the rule – *to behead the victim*. It's all in the handbook – I'll show it to you if you don't believe me – Satan-worshippers chop the heads off anyone who denies their god – the Devil. That's how it goes, Art.'

Arthur Fling slumped back in his chair and tried to outstare his tall subordinate on the other side of the desk. It didn't work. Terry Lenning didn't flinch from the survey. Conviction gave him courage.

'What else have you got?' Fling asked very softly, 'that makes it a master plot of some kind – that makes it *organized*?'

For answer, Lenning nudged the stack of charts and lists across the desk in Fling's direction. They rustled like leaves in a high wind.

'Look those over when you have the time. You'll find that Betty Taylor, Renee Richardson, Sonia Black, and Arlene Busby were each independently wealthy. One was a Boston heiress, another was the widow of a California oil tycoon, Black was a vocalist with three gold records, and Busby owned her own clothing store on upper Broadway. That means four wills are now being processed – there's something like three and a half million dollars involved – and none of them had any next of kin. In fact, they didn't handle their lives very smart at all. Black and Busby never got around to making out their wills and Taylor and Richardson leave all their loot to their estates – and the Government.'

'Then,' Arthur Fling smiled tiredly, 'you are indicating a lack of motive altogether. Where's the profit in these murders? Nobody benefits, according to you, but the United States Government! Or are you by any chance suggesting the IRS has found a new way to add to the national budget? And wouldn't that be a kick in the head?'

Terry Lenning shook his head, smiling bleakly.

'No claims have been made yet. Nobody's stepped forward.

22

But I'm waiting, don't worry. Still – profit isn't the only motive when it comes to these series-type murders. You know that as well as I do. Manson didn't kill all those people for money; Kurten certainly didn't, nor did Jack the Ripper or the guy who murdered all those nurses – '

'Terry, Terry,' Fling sighed again, 'stop beating around the bush. Now what are you trying to tell me?'

'Devil-worship,' Terry Lenning repeated. 'Taylor, Richardson, Black, and Busby were killed by someone who *believes in the Devil*. That's motive enough, you know, if you really buy that scene.'

'The Devil made me do it, eh?' Fling growled unhappily.

'Something like that,' Lenning agreed.

And there they both left the matter, until the next headless corpse came bleeding into their lives and their daily routine.

Which wasn't very far off.

Twenty-four hours away, in fact.

But later that day, a strange dinner engagement would mark the entrance of Philip St George into the bizarre affair.

Philip St George – the man they called the Satan Sleuth.

DO YOU BELIEVE IN SORCERY?

Because Norma Carlson was a newspaperwoman, feature writer for the *New York Times* actually, and her phone call had come at Philip St George literally out of the blue, he saw no reason to question her choice of their meeting place for a meal. After all, when a woman phones you and offers fare at one of the best restaurants in town, the *Wallingford* in Rockefeller Plaza, and also happens to be tall, willowy, and Nordically blue-eyed, bronzed-skinned, and yellow-haired, it would be ridiculous to refuse. Once Philip St George was certain that Norma was not sniffing around for an interview, for he never gave one to anybody since the tragedy up at Lake Placid, and he heard the urgency and worry in her too casual and striving-for-flipness telephone talk, he rapidly agreed. Norma needed him for something; he sensed that the way he sensed all things. Spiritually, acutely, sensitively. The lady was in trouble of some kind and if she thought Philip St George could help her, well, why not?

He liked her, rather admired her. He always had since those terrible first few weeks after Dorothea had been butchered by those bastards up there in the desolate country estate – Norma Carlson had not been maudlin, sob-sisterish, or seeking sensation when she'd written so feelingly and tastefully, and compassionately, God bless her, about the tragedy which had shocked a generation. The millionaire adventurer and the Miss America wife – victims of a violence-ridden universe – Philip St George had never forgotten her reportorial dignity in the face of the news event of the decade. Still, Norma, like all the rest of the world, could not know what had become of Philip

St George since the ordeal of '73. Or could she? That was another interesting reason for having lunch with Norma Carlson. Besides her Nordic attractiveness.

It couldn't be love or sex which had led to the call.

Philip St George had hardly looked at another woman since the day Dorothea Daley St George had died. Passion had died, too.

He had been too busy – with other things.

His own very private crusade, the secret career which only Sidney Kite, his lawyer, could know for sure. And even Sidney was finding it difficult to credit his own suspicions in the face of the staggering exploits being recorded all about the hectic globe by a mystery man who was behaving like the Lone Ranger on those old radio serials.

Coming, doing his good deed, then vanishing without waiting for thank you's or rewards of any kind. An incredible, avenging philanthropist.

But, of course, Philip St George didn't need any money, did he?

He was the tenth richest man in the country and, according to Dun and Bradstreet, was moving up to the ninth position thanks to the increasing profits of the St George restaurant chain. Phil was a regular tycoon!

Yet what did a daredevil millionaire explorer, still in his early thirties, do with all his free time? A man who, like the young Alexander the Great, had no more modern worlds to conquer? Before Dorothea's death he had already filled some three pages of the National Geographic Society's *Who's Who* with stunning accomplishments.

He had scaled Everest, climbed the Matterhorn, whipped the North Pole, dug up archaeological finds of staggering importance, scouted the Marianas Trench in a private nuclear submarine, explored with Cousteau, called Tensing by his first name, and had a room full of Explorers Club medallions and awards – but had he stopped living altogether? Or was this cockamamie crusade, as Sidney Kite christened it once in his Broadway law offices, this Satan Sleuth business, for real?

No one could safely say. Not the FBI, CIA, nor Supreme Court.

25

No one really knew. Less was known about him than about Howard Hughes.

Unless Norma Carlson, with her newspaperwoman's unerring instinct and talent for the truth, had stumbled onto something, finally.

Again, perhaps the best possible reason for their reunion.

If one needed reasons for a beautiful woman's company.

Philip St George decided she was far more beautiful than even he had remembered as he found her sitting in a quiet booth in the corner furthest from the front door of the Wallingford, removed from the dining mob.

She seemed blonder, more blue-eyed, bronzed, and willowy than ever. The wide picture hat and the sleeveless aquamarine wool dress clung to her length like a second skin.

'Phil,' she greeted him, her smile brilliant.

'Norma.'

Their hands touched, squeezed, and he could tell that she was afraid. The tremor of her fingers was unmistakable. He smiled warmly to relieve whatever tension she might be operating under. It was far too soon to start firing questions at her. She had to take the lead. No matter what. He sensed that she had chosen this remote booth for its privacy and distance from the din and clamor of the main dining room. Away from oglers.

'Wallingford,' he teased. 'You must have gotten a raise in salary since last we met. Steaks here start at ten dollars.'

'I pawned my Pulitzer.' Her eyes were hoping for gaiety but he could see the shadows in their pools. 'Never mind. I called you, the lady's buying. Even if you are the richest live one for miles around, and – oh, how have you been, Phil?'

'Tolerable. And you?'

'Tolerable's good. Serviceable word, that.'

'Why aren't you drinking?'

'Just arrived about five minutes before you. I've ordered for both of us. You still drink Gibsons, I hope.'

'Right on. And if memory serves me, you had an absolute passion for that drink that rusts the lining of the stomachs of beautiful women. Straight rye. With a beer chaser. You don't look like a newspaperwoman, Norma, but you certainly drink like a newspaperman. Boilermakers. Ugh.'

She laughed at that and her stellar, first-class teeth flashed. A smile on Norma Carlson's face was like a sunset in the Arctic. Absolutely dazzling. Philip St George lounged back against his cushions.

'I was surprised to hear from you. Flattered like hell – but surprised. You know what I mean?'

She shook her head and the dazzling smile was touched with sadness. And no little amount of disbelief. The blue eyes misted a little.

'You're a weird one, Phil. You know that?'

'I'm willing to listen to you,' he hedged. 'Tell me about that.'

'Well, for openers, apart from your filthy-rich background and your fame, you make Robert Redford look like Huckleberry Finn. You've got every female in this place suffering whiplash because they're all craning their necks to get another look at you. And here I am with the whole pack of them wondering what I am to you in the scheme of things. So what's surprising about me calling you up and wanting you all to myself at dinner?' Norma Carlson broke off, almost apologetically. 'Oh, Philip St George. You *are* impossible. So dashing, so romantic, so unconceited. Where's your ego, man? Where's the lust in you?'

He leaned forward suddenly, reaching, taking both her hands in his. Before he could speak a silent, somber waiter came with their two drinks. He left menus but Philip St George ignored them. He stared into Norma Carlson's eyes with slow, deliberate intensity. When he at last spoke it was in a voice that did not carry further than her ears.

'Norma, I'm here because I'm interested in what you want to tell me. And you have brought me here to tell me something. Not because you are interested in a roll in the hay. Okay? Don't play the fool with me. It's not necessary. I'm your friend. I will help you if you need me. If I can. It's as simple as that. I'm not going to go on a fishing expedition with you to get you to open up. If you've had second thoughts, all right about that, too. We'll both order our steaks – the *entrecote* here is superb – rap about old times, and let it go at that. So you call the play and that's how it will be. Fair enough?'

Norma Carlson's lovely face seemed to drain of all blood.

It was an illusion, of course, but Philip St George could see the almost electrical effect his quiet little speech had had. Norma was fairly collapsing in his hands, the feeling leaving the fingers he clasped across the table. As if all life and animation had fled. As if a rag doll sat lifelessly before him. Coming undone.

'Christ, Phil, you *are* psychic. They said you'd gotten that way – they say so many things about you now. . . . '

'Skip that for the time being. Let's stay with you.'

'Phil – it's so nightmarish – I don't know how to lead into it – isn't that a scream? Me – I feel like I'm caught in some surrealistic fun house. I'm awake all the time but I'm dreaming – I must be!'

'This day and age is no picnic, Norma. Which nightmare are you talking about in particular? Watergate, energy crisis, Arab-Israeli War, or famine? You tell me, Norma.'

'Phil, Phil . . . ' She whispered his name now, her eyes poignant with despair and helplessness in the subdued light of the intimate booth. 'I'm – mixed up in a coven – and I can't get out – and I just don't know how to handle it. I don't know what to do – me, Nifty Norma, fearless girl reporter. . . . '

Philip St George's face was expressionless. Only his eyes stirred. Norma's voice had trailed off into wretched, personal misery.

'Drink up,' Philip St George said calmly, 'and tell me about it. From the top. Without editorializing.'

So Norma Carlson did.

'Do you believe in sorcery, Phil?' she asked before she began her strange tale, draining a stinging shot of rye.

'Do you?' Philip St George replied without inflection.

'I didn't,' Norma Carlson shuddered in her second-skin woolknit aquamarine dress, 'until I met a woman called Catherine Copely.'

That, in effect, was the beginning of Philip St George's enmeshment in the affairs of the witch known to her coven as Sister Sorrow.

A woman, for whom Death seemed to be a handservant.

THE CARLSON BEWITCHMENT

'I fell in love,' Norma Carlson said in a low, unhurried voice, somehow in control of herself despite whatever she was about to relate, because of the presence of Philip St George. 'You would have liked him, I think. Paul Tatum. A *Life* photographer with brilliant credits. I met him in Palestine last summer. I'd gone there to cover Henry K.'s trip for the magazine. Paul was on the spot, taking pictures of the Golan Heights mess. Well, it was a whirlwind thing, all right. I'd never been so in love before. Do I have to explain that to you? No, I don't suppose I do. At any rate, besides his lover boy-dream man good looks, he was a fine person. A wonderful way of looking at things – why, he even understood the needs of a woman like me; why I was so independent, so much my own woman. I loved him, Phil. Really loved him – and when I had to come back home and he had to stay on in Israel, the plan was to get together this fall and think about a permanent hitching-up.'

Norma Carlson stared down at the shot glass in her slender fingers, twirling it very slowly. The glass was empty.

'A mortar shell wiped him out on September tenth at five o'clock in the afternoon while he was canning some film for Air-Express. I got the time, the place, and the circumstances from a fellow correspondent who was on the scene and got out of it alive. I only mention that because at that precise moment I was in Bonwit Teller's picking out sort of a trousseau. Screwy world, isn't it?'

Philip St George said nothing, his eyes never leaving her face.

'Well – I went to pieces, too. Cracked up. Couldn't work,

29

couldn't write a line, crying all the time – they gave me a month's leave of absence and told me to go to Cannes or Paris and get myself interested in some physical love again. Something like that. I couldn't do it – I don't know. Paul and I had only had a few weeks together but he had become some kind of answer for me. You know me, my career – I'd been Little Miss Independent long before Women's Lib and Steinem. So there I was, in a daze, not caring how I looked or how I ate or what I did when I met a man.'

'A man,' murmured Philip St George, just to keep her down to earth.

'Yes.' Beautiful Norma Carlson smiled. A ghost of a sad smile. 'Duane Farmer. A news commentator. Does world comment for CBS now and then. Not a big wheel but a fellow journalist of sorts. It wasn't what you might think. Farmer is over fifty, as ugly as sin, and definitely not the antidote for a dead love. But he did offer something else. Something I was totally unprepared for, and never would have believed in a million years I would go for. My defenses were down, of course – I was Poor Little Nell, alone and lost. I listened to what Duane Farmer had to say, and in the end I went with him.'

Philip St George silently shooed away a polite waiter who had materialized to take their orders. Norma Carlson was too intent on her story to notice. She sipped her beer, slowly, as if to savor it.

'Farmer took me to a penthouse overlooking Central Park. The home of a woman named Catharine Copely. The strangest, damnedest, most beautiful and bewitching female I have ever seen in a decade of the Beautiful People, the Jet Set, the – you name it. And there, with a lot of other odd people, queer characters, I was introduced to my very first legitimate, flesh-and-blood witch. What I mean – the real McCoy. I thought I'd been around, knew the score, but she sold me a bill of goods. All the way.'

'Catharine Copely is a witch?' Philip St George asked softly.

'If she isn't, the breed doesn't exist,' Norma Carlson breathed with sudden, fierce intensity. 'She opened my eyes, she indoctrinated me – she told me she could bring Paul back from the dead so I could talk to him – see him – touch him – see his crazy smile again – hear his laughter – and as God is my judge, I bought it. The whole thing. Me – Pulitzer-Prize-winning Carlson! The darling of the you-can't-pull-the-wool-over-my-eyes league. I honestly don't know how to make this sound logical or even half reasonable, Phil. But it's that woman – Catharine – her eyes, her voice – '

'Easy, now. I'm still listening.'

'I know you are. But I can hear what I'm saying, what I'm trying to tell you, and I'm choking on it – oh, Phil – ' She stared at him very earnestly across the table. 'What would you think of me or say if I told you what I've been doing these last few weeks? At midnight – up there in that hell-hole of a penthouse? How would you feel knowing that the woman you're sitting with has participated – *willingly, undrugged, unhypnotized*, in the Black Mass? That she's run naked in a pentagram circle – performed every kind of sexual perversion with a half a dozen men, under the light of a full moon – that she's become one of the Devil Circle who worships the Dark Powers? Who blasphemes the cross and Christianity – who, who – ' She broke off, her voice rising, knowing her voice had gone up, aware only of Philip St George's firm fingers boring into the palms of her hands to steady her. She stifled a low sob, unable to continue, incapable of looking into the handsome face with the sober, piercing eyes directly before her.

'Look at me,' Philip St George commanded in a quiet voice. 'Don't withdraw now – not when you have finally come out of that place you've been hiding in. Look at me, Norma.'

Slowly, almost involuntarily, her eyes came up to his.

'I'm not here to judge or decide or dole out forty lashes,' Philip St George said evenly. 'I came to help you. What do you want me to do?'

Norma Carlson stared at him. Utter disbelief wiped away the former expression of self-loathing and disgust. And terror.

'You mean – you're not shocked – or revolted – with me?'

'I mean I'm your friend, Norma.'

'You're also – wonderful,' she muttered almost inaudibly. 'Phil, Phil – she's never shown me Paul Tatum – she promised that for later – she's a witch though – God, is she ever! She

31

can transform people into animals – I've seen that with my own eyes, and it's no trick, no sleight-of-hand. She's for real, Phil! And she is *evil. Really evil!* And now I'm frightened. I want to get out – quit the coven – but things have happened to anybody who has tried – awful things – like the other night.'

'Yes?' It was the barest prod, a form of goad, to keep Norma Carlson talking, giving vital information, before her courage ran out.

'Carmody – a dress manufacturer – the poor bastard didn't have any chance at all – not against Sister Sorrow – *damn her black soul!'*

'Sister Sorrow?'

'That's her cult name. A title. It's part of the whole program. And the other night when he wanted out, she just looked at him with those eyes of hers, like she was making a wish. And, God, she changed him into a snake – *so help me God!* I saw it, with my own two eyes.'

Philip St George, if he could consider such a miracle possible at all, did not by word or facial expression indicate his opinion of any such unlikelihood and impossibility. Rather, he kept his face immobile. Poker-blank. Norma Carlson, despite her state of near hysteria, had not lost all of her native shrewdness, however.

'You think I'm crazy, don't you, Phil? Gone off the rails.'

'I didn't say that.'

'You must be thinking it but you're too goddammed polite to say so.'

'I want to hear the rest of what you have to say. Then we'll both decide what to do about Sister Sorrow and the coven. I don't deny the Devil, Norma. I simply don't believe in his actuality.'

'If you think a thing,' she whispered, eyes almost glazed, 'that makes it so, doesn't it?'

'I'll postulate you another, then. Thinking a thing is so doesn't make it so.'

'Oh, Phil. Here I am babbling away with all this rot and we haven't ordered and now I feel real foolish. In the daytime it all seems so insane and unreal. But then the dark comes

32

around, and it's all too vivid. All too true. Does that make any sense at all?'

Philip St George nodded. 'Shadow and substance, Norma. But no more of proverbs and aphorisms. You need help, I'm prepared to help you. I take it there is another meeting of this coven and you don't want to go and you're afraid of what this Sister Sorrow will do. I understand that. And you have come to me. May I ask – and leave out the fancy ribbons – why me?'

'Why not you?'

The dazzling smile she surrounded that riposte with was very nearly the old Norma Carlson. A good sign she was settling down a little from that high plateau of fear and fantasy. Stabilizing her nerves.

'Answer the question. I'm interviewing you now.'

'Phil,' suddenly her tone was very serious and even more guarded, 'I'm a newspaperwoman. I know a lot about you. I'm not an idiot who can't read between the lines. I know what goes on in the world. I hear things, too. Like the Fletcherville Werewolf story. And Port Elizabeth and that voodoo-cult thing. And that exorcist tale from Cape Cod. I know there's somebody running around trying to stamp out the Devil all over the globe. A man who comes and goes like a ghost, doing all his good deeds and then taking a powder. Nobody knows who he is, what he is, or why he's doing it. I can think of only one person good enough to do all those things. One person with the background, the money necessary for such unpaid-for adventures. A man who saw his own wife killed by cultists and lived on to do something about that, too. And then kept on doing it. A guy like you who had done it all – mountain-climbing, exploring, daredevil stuff – the whole bit. Philip St George, if you're not the mystery man everybody calls the Satan Sleuth then I'm not in this room, you're in Tibet, and Catharine Copely is a leftover from Barnum and Bailey's – which she isn't, God help us all!'

Philip St George did not reply immediately. His own bronzed face was impassive, only his striking eyes glinting. Searchingly.

'When is the next meeting of the coven?'

Norma Carlson shivered visibly. Her shapely shoulders shook.

3

'Even telling you is a sin. A crime against the Circle of Truth.'

'But you are going to tell me, aren't you?'

'Yes, of course.' She nodded rapidly, as if speed would solve the dilemma of indecision to rebel against Satan, the forces of Darkness, and Sister Sorrow. 'It's the penthouse again, as usual, and that's located – '

Later, when he would look back and try to remember exactly what had happened at the precise moment when Norma Carlson was telling him about the meeting, even Philip St George would be hard pressed to explain what had occurred. Not only *what* had happened, the incident itself, but in fact the true order and sequence of the events which had shifted so alarmingly.

Memory and vision, regulated as they are by the mind of the person who is eyewitness to tragedy and startling phenomenon, can never be wholly accurate and reliable. Not even when the observer is a Philip St George who has seen more things in this life than most men will in a century of living. Always has that been so.

Never more so than now.

With Norma Carlson bitterly torrenting out the words which must have, inevitably, sealed her fate. Doomed her to the grave.

Wallingford, buzzing with sounds familiar and indigenous to a popular restaurant in midtown Manhattan, would never be the same again. Not quite. Table-talk hum, culinary clatter, the drone of subdued music filtering under the scene, like the score for a movie, all of those sounds vanished and disintegrated within a few seconds. The few seconds it took Norma Carlson to suddenly stop talking, a quizzical, almost childlike expression of bewilderment passing over her lovely face – and die.

Death, whatever it was, was instantaneous.

Death came to Norma Carlson, at the age of thirty-two, at the peak of her reportorial powers and the nadir of her personal life.

Between those disparate polarities lay only enigma, enormous confusion, and a great, deep, and abiding terror.

And horror, too.

On the phrase 'that's located – ' she abruptly put her red

34

lips together, her eyes shot open in gaping panic, her throat muscles leaped into focus like piano cords, and the palms of both hands slammed down on the tabletop, sending the beer glass spinning, sloshing amber contents over her wool-knit dress. Philip St George, stunned for an all-important second, froze where he sat and then his reflexes responded with animal coordination. He reared from his seat, reaching across to catch Norma Carlson before she sagged back to the cushions. The picture hat's wide, circular brim, shadowing the lovely Nordic face, prevented immediate assessment of the situation. Other than the simple fact that the girl was in trouble. Bad trouble. And then it would have been too late no matter what he did. As the postmortem later would prove all to indelibly.

An eerie scream charged out of Norma Carlson's throat.

A scream that blended into a strangled, choked gurgle of unintelligible sound. Of discordant, bubbling agony.

She slumped, stiffening, arms and legs thrusting violently, kicking, thrashing, and not even the powerful young man with her, at her side, could do anything about that. The darkening crimson, mottled hue of the perfect blonde complexion was telltale in itself. Grisly and plaguelike.

The gruesome protrusion of the pink tongue was the final reality. After that, there was nothing.

Nothing except disorder and uproar in Wallingford, dismay and despair in the soul of Philip St George. Satan had touched him once more.

Norma Carlson had died sitting across from him.

She had been poisoned.

That was simple enough.

What wasn't simple at all was how the poison had been inducted into her system.

In itself, that was the greatest magic of all. The topmost puzzle. The number-one riddle.

For the time being, it was nothing short of *witchcraft*.

The Devil's kind.

FLY A KITE

'Did you love her, Phil? Is that it?'

'No, Sidney, I didn't love her. She was just a beautiful human being.'

'Lots of beautiful human beings die every day. In wars, to cancer, in accidents – why are you so upset?'

'The police were very stupid about the whole thing. I thought the third-degree technique went out with Cagney movies.'

'Don't believe it. You're a rich one, Phil. A very rich man. Cops making in a whole year what you could spend in one day without feeling it can cause a lot of resentment. You know that.'

'I don't know anything sometimes. Like just how did Norma die? The cyanide wasn't in the beer or the rye. She didn't even touch a glass of water. Or smoke a cigarette. We met, talked for about twenty minutes, and then she keeled over. Right before my eyes. She was dead before she cleared her throat.'

'Phil, look at me! Stop staring out that damn window. You've seen that view a million times. Broadway hasn't changed, it never will no matter how the cockamamie real-estate boys go hogwild playing Monopoly. We got to talk this thing out. You're a material witness if not a first-class suspect and ain't that a kick in the head?'

Philip St George turned easily from the broad picture window which was the only outstanding feature of Sidney Kite's *sanctum sanctorum* in the trim, tenth-floor complex known as Kite, Dorn, and Schindler.

It was the morning of a new day, winter grays and blues painting the Manhattan skies. Young Phil, resplendent and

36

handsome as ever in charcoal-black suit with a neat splash of carmine tie, hardly looked like a man who had spent most of the night before in a police station, answering questions and trying to explain to a not too friendly Homicide detective why a famous newspaperwoman should be poisoned by 'person or persons unknown' while merely dining in the ritzy precincts of Wallingford. A hurried phone call to Sidney Kite had released the prominent Philip St George on a writ of *habeas corpus*, but as matters stood, Phil was still in plenty of hot water.

If anyone could bail him out, Sidney Kite was the lawyer to do the job. Frenetic but shrewd, overweight but clear-minded, affectionate as a father and brilliant as a defender, the middle-aged, lifelong friend of the St George dynasty was anxious to know everything there was to know.

Despite Phil's usual, cool injunction: 'Don't worry about it.'

Sidney Kite had always worried. Never more so than since Dorothea Daley St George had been murdered and her husband had gone off on his cockamamie crusade. Trying to save the world from the Devil . . . Talk about thankless jobs – with no pay, yet! Not a red cent.

'Phil.' Sidney Kite lowered his growl a notch and tried to smile. Behind the pedestal of his cluttered desk, which brassy, sophisticated Miss Walters usually kept a shipshape order, he seemed a dough-faced, large-nosed man with little to recommend him. Appearances were all too deceiving. What Sidney Kite did not know about law would not be worth knowing. 'I need some more facts. All the facts. What about this cyanide? The autopsy must have shown something. Some manner of entry.'

'Nothing,' Phil said flatly, no wonderment in his tone. 'I have to hand it to them. They tried everything. Checked Norma Carlson's mouth for a recent filling with a gelatin cap which could have dissolved while she was talking to me. Her lipstick proved to be zero, too. But that poison was in her bloodstream. And it killed her. And they don't know how it got there. No skin punctures – nothing on the roof of her mouth or tongue. Not a trace of how the stuff entered. They ran me through the dry cleaner, too. They had to. I put up with it because I wanted a clean bill of health as far as the police are concerned.' He took a deep breath and his eyes glinted oddly. 'Norma Carlson was killed by a concentrated dose of cyanide, normally a fast-acting poison, and nobody in the police lab can say how it got into her. And I can't either. It's – *magic*.'

'Magic?' Kite echoed, sniffing the air of the room in disgust. 'You going to start this Satan stuff again? Come on, Phil, have a heart. There has to be a logical explanation. Something for a courtroom.'

'Well, there isn't. Not at the moment, Sidney.'

Kite shook his head, eyes narrowing. 'What did she want to see you for in the first place? Was she husband-hunting?'

'What makes you say that?'

'I read about that cameraman she went ape over in Israel. And all about his getting killed and her taking it hard. I knew you always liked her because of her good taste when – Lake placid happened – and I also knew she knew that too. And why not? She would have been good for you. As good as any I've seen lately. Poor *schicksa*. A real waste, getting dead like that. But, jeezis, Phil – '

'Yes, Sidney?' Philip St George might have been miles away.

'Why? Who'd want to do her in? With cyanide yet? And a magic act if what you say is true.'

'I don't know. Not right now at any rate.'

'What did she want to see you about – if she wasn't shopping?'

'Nothing that would interest you, Sidney.'

'You're lying to me, aren't you?'

'Yes. A little, I'm sorry to say.'

'And you're not going to tell me, either?'

'That's right. I'm not going to tell you.'

'*Schlemiel*. I could be a big help. But no – Lone Ranger stuff, is that it? Phil, have a heart. You go running off again and my ulcer does the *hora*. You know that. Let me in on it, will you? Have some respect for a man old enough to be your father! Show me some consideration.'

'I'm being kinder to you than you know by not involving you in this at all. Stick to law, Sidney. I have to do my own thing.'

'Which now is?' sighed Mr Sidney Kite, helplessly, recognizing the old familiar pattern of their past relationship. The mixture as before.

Philip St George strode to the desk and stared down at Sidney Kite.

'I'll need your help again. A great deal of it this time. More information and paperwork than I've ever given you before. Or asked for. For once, I'm not poking my neck in, you see. You said so yourself. I'm already involved. A material witness in a murder. At least you have to admit that anything I do this time will be helping clear myself with the police. How would it look – the head of St George, Incorporated with the suspicion of homicide hanging over his fair head?'

'Cut the comedy,' Kite grunted, for he had already yielded, knowing that he couldn't do anything else. He pyramided his hands and glared moodily from their apex. 'So what can I do for you this time?'

Philip St George's penetrating eyes slitted thoughtfully. He suddenly placed a hand inside the inner breast pocket of the stylish suit and withdrew a folded sheet of letter-size paper which he handed across the desk to a frowning Sidney Kite. With a client like Phil, the middle-aged lawyer never knew what to expect.

'There's a list of names,' Phil said evenly enough, 'which I want you to track down for me. They shouldn't come as too much of a surprise to you considering all the newspaper space they've been given of late. The beheaded corpses which have been found all over our fair city these past weeks. Ring a bell?'

Kite had unfolded the crisp white sheet of regulation bond, rapidly found the four neatly typed lines rendered one below the other, double spaced. And that was all. Nothing more, nothing less. Just four names.

'Betty Taylor, Renee Richardson, Sonia Black, Arlene Busby,' he read aloud like a schoolboy reciting a lesson. 'That's all you want for now? No special purchases, like hand grenades, smoke bombs, or carton of technical equipment? Too easy this time, Phil.'

'Not so easy, Sidney. I want to know all there is to know about those four women. Dating back to their first lollipop. It should keep Miss Walters and you busy all week.'

'We can handle it. What's so special about these dead ladies, Phil? Isn't it enough for the cops to handle?'

'Do you recall anything at all about the crimes?'

Kite made a sour face and abruptly looked dyspeptic.

'Lawyers read newspapers no matter what you might think. And who could forget slaughters like those? Naked, heads chopped off, mutilation. It's been a panic, to hear the police department tell it. But, Phil – if I get your drift, you're hinting this is another one of those cockamamie Devil-worship things and maybe it ties in with your Carlson lady. Am I getting warm or do you play dumb again?'

For answer, Philip St George was already on his way to the door.

'I'd appreciate any speed you can manage, Sidney. I'll be home when you have the material. If not, leave a message with the answering service. Thanks for your time, Sidney.'

'Oh, it's a pleasure, believe me,' Sidney Kite snorted, unable to ever really be angry with his young client. 'There's nothing I like better than operating in the dark. Know what I mean?'

The fleeting ghost of a smile flitted across Philip St George's handsome face. On the threshold of the doorway, with his faultless tailoring, superbly elegant figure, and uncommonly handsome head, the distinguished streaks of silver showing at his temples, it was impossible to imagine him being anything other than what he was: scion to a family fortune, dashing explorer, idle rich boy with nothing more strenuous to do than read the *Wall Street Journal* to check his countless holdings.

'Mr Kite,' Philip St George said banteringly.

'Yes, Mr St George,' Kite rumbled, getting into the spirit of things. 'You going to leave me laughing with a funny exit line?'

'No. I was merely going to say, for the record, to offer you some shred of information so that you will not be totally ignorant of my activities, that I shall for the next few days be engaged in what is euphemistically referred to as a *hunt*.'

'Hunt?' Kite didn't like the ominous sound of the word. In fact, it sent an involuntary chill down his jaded spine. 'What kind of a hunt, Phil?'

'A witch hunt,' Philip St George replied without hesitation, and left the room. He closed the door very quietly so that it seemed like the echo of the phrase lingered behind him. Sidney

Kite blinked, growled again, shook his head in despair, glanced at the typewritten sheet lying on the cluttered desk before him, and finally raised his eyes to the ceiling in fervent appeal to some kind of god. Any divine being.

Inevitably he surrendered altogether and flicked on the inter-com box positioned close to his right hand. His doughy face was now contorted into a frieze of tiny angles and lines. Almost of pain.

'Yes, Mr Kite?' Miss Walters' smooth, professional voice filtered from the black box. Walters was a fixed star in the Kite scheme of things. Kite, Dorn, and Schindler had never had a finer secretary-in-chief. Walters was a paradox. A looker who could really take shorthand, run an office, and keep a law firm on an even keel. Genuinely intelligent.

'Irene,' Sidney Kite muttered wearily, 'come in here, will you? And bring your steno pad. Young Galahad is making his move again.'

'Right,' Miss Walters said. 'I'll bring the Bromo, too.'

Though she knew absolutely nothing about Philip St George's private crusade, Miss Walters had been around long enough to know the effect of the young millionaire's infrequent visits to the office.

All too well.

Sidney Kite's ulcer would start acting up once more.

And she, Walters, would be busier than a one-armed paper-hanger or a woman with ten kids, doing all kinds of legwork, research, and file-digging in such diverse places as the public library, the *New York Times* backnumber morgue, or anywhere else Philip St George might have in mind.

Miss Walters, who possessed a *Vogue* figure with a glossy face and spohisticated sheen to match, had once set her blonde head for Young Moneybags after his wife had been killed. But even waiting a respectable interval of time had not helped, nor had coquetry, tight-fitting sheath dresses, and every variety of dropped hints, cajolery, and feminine strategy. Philip St George always looked right through her whenever he appeared, as if by magic, at the office. Irene Walters got the message and stopped trying. Single and in her early thirties, she did not want for male companions. Nor was she interested in marriage, either.

Unless a comfortable, lavish existence went with it.

Still, Walters was woman enough to lament the utter and tragic waste of St George's phenomenal assets, such as good looks, millions, and a splendid physique. The dame that finally nailed the man would be in clover all the rest of her life. Irene Walters was sure of that.

What she wasn't sure of, nor was anyone else that winter day, was the fifth victim was already lying in the cold, barren basement of a hovel on Minetta Street in the Village. Dead for hours.

Where she would be found by truant schoolboys playing games in the dungeonlike atmosphere of the old cellar beneath an empty, crumbling, deserted three-story building. Schoolboys faced with their first ugliness.

Schoolboys who would come screaming up out of the darkness because they had just found a naked woman, all bloody and horrible, whose head was squatting on the bleak stone floor, just a few feet from the outraged body. A head propped up, with its eyes still staring glassily, as if wondering what it was doing separated from the figure is belonged to.

The fifth headless woman would prove to be a run-away daughter from a Frisco clan of bluebloods, one Elvira Davis, age twenty-one, heiress to a brewery fortune, whose parents had perished in a jet disaster over South America the year before. Elvira had rebelled and run.

But all of that would come out later. Much, much later.

After the police arrived to take over the official investigation.

An investigation in which they were already down-right baffled.

And getting nowhere with, according to every newspaper in town.

And a growing awareness of the sudden and terrifying gruesome wave of murders was sweeping the city. The Black Mass aspect was all too clear.

There had to be a connection between five corpses without heads.

What Terry Lenning had tried to convince Arthur Fling

42

about at the FBI Headquarters office in Manhattan, he would not have had to do a selling job of on Philip St George at all. Not in the least.

The Satan Sleuth had not the slightest doubt about the five murders.

It smacked of cultism of some kind, Satan worship, from the word go.

As such, it was but another case for the Satan Sleuth.

Philip St George had not cared to mention the name of Catherine Copely to Sidney Kite. There was no need, somehow. He intended to investigate Miss Copely himself. And Duane Farmer. Some things were best left to one's own pursuit. Especially witches. *Real* witches, if such things could be. If that's what Catharine Copely truly was.

The Satan Sleuth was determined to find out, on his own, his own way. He was going to chose his own weapons, also. It was hunter's choice.

If there were any connection between five headless female corpses and a mysterious coven somewhere in a penthouse overlooking Central Park, well, he would see to that, too. South, east, or west, he would locate that aerie in the sky. Norma Carlson would have wanted him to.

There couldn't be any doubt about that, either.

Not to a man like Philip St George.

He would not look north of the mighty park for the coven.

Harlem had enough trouble and woe without adding covens to the score. Norma Carlson had not mentioned the word *black* – and a white witch would not ride her gold-plated broomstick anywhere near Lenox Avenue.

CATHARINE COPELY,
GODDESS OR FIEND

She would have been judged beautiful by any standards in any century. Her skin was flawless. Pure ivory whose faintest tinge of color was merely heightened and accented by incredibly lustrous dark eyes showing beneath two perfect black streaks of brow. For a nose and a mouth and cheekbones and jaw, an unknown chisel had arranged composition and order with incredible arithmetical skill. With all this matchless quality bordered and somehow crowned, topped off as it were, by full-length, flowing, raven-hued hair, it was small wonder that any male within talking distance of Catharine Copely fell immediately under her spell. The label *enchantress* was not too far-fetched when applied to the titular head of Chic, Incorporated, the successful women's fashion boutique on upper Fifth Avenue.

It was Catharine Copely's unique talent to instantly strike any man who met her as, perhaps, the most beautiful woman in the world. Women, friends and foes alike, as well as perfect strangers, would forever turn their heads and stare when Catharine Copely drifted into their presence. *Drifted* is the exact word. Catharine Copely did not walk, she glided, shimmered, and the strange ambience of her person had everything to do with her rare beauty, her uncommonly tall and slender figure, and the seemingly endless string of saris, caftans, and robelike garments she obviously preferred, day and night. At the office or at home. Catharine Copely had never been a movie star or an actress or a model of any degree, nor did she possess a wealthy, titled background history, and no one could remember where she came from or when she suddenly burst on the New York scene. Suddenly, she was *there*.

And that was enough.

No one asked questions; no one really cared.

It was sufficient that this bright and luminous star had abruptly descended into the midst of people's lives and careers. Like a *happening*.

In Manhattan the only credentials one really needs are position and power. And personality. Whatever that position, power, and personality actually are, is secondary.

Catharine Copley was *now*, *contemporary*, and *in*.

She *was*. She had become an integral fixture in New York society.

So, monied and famous citizens bought clothes from her, lionized her and gloried in her beauty, feeling favored by her charm and interest.

Even though they didn't know a single blessed thing about her.

Except perhaps Mr Peter Bond, who appeared in her Fifth Avenue kingdom on the chill morning following the discovery of Elvira Davis's headless corpse in the bleak basement on Minetta Street in the Village. The morning headlines howled at the authorities, screaming for a rapid end to the series of weird slaughters filling the daily papers, but none of that panic or disorder was visible in the ultramod, cleverly subdued environs of Chic, Incorporated. Though the horrible affair was very much on the mind of Peter Bond. As was Catharine Copely.

Not even Sidney Kite or Dorothea Daley St George, had she still lived to love him, would have recognized Philip St George in his perennial disguise of Peter Bond. For Phil had given his favorite bogus characterization yet another turn of the screw. Since assuming the crusade and arming himself with every weapon at his disposal for the war against Satan, Sidney Kite's most troublesome client had become a veritable Olivier with a makeup box. And while he was waiting for the Kite-Walters combination to hand over a substantial file rundown on the cult murders sweeping the city, he had decided to invade the enemy camp. In a direct frontal assault.

Norma Carlson had given him the lady's name. Easy enough to look up, simple enough to track down. What wasn't so easy nor so simple was how exactly to attack the problem. Hence, this head-on confrontation.

It would have taken a real witch to spot the handsome Philip St George in the covering façade of Peter Bond. Dream man was *kaput*.

The height of six feet was lessened by a slightly stooping gait. Gone was the bronzed complexion which had known the Arctic and all forms of high adventure. Peter Bond seemed as if he hardly ever saw the sun at all. The Cary Grant shock of curling black hair with silver streaks was now a dull, muddy brown, fairly unkempt, with a tendency to resist any kind of combing. The sensitive Apollo face was hidden behind lumpy cheekbones. A strong, semiaquiline nose threatened the world at large. Brown contact lenses masked the Polar blue eyes. The impeccable St George tailoring now showed a rather baggy, not too well fitting brown tweed suit.

Imagine Robert Redford transformed to George Segal.

This was the man whose appearance greeted Catharine Copely as a genial, courteous shop assistant ushered Peter Bond into her inner office.

Catharine Copely looked up from some preliminary sketches she was making of a belted pants suit and stared at Peter Bond, charcoal poised.

Peter Bond stared back, smiling uneasily as a Peter Bond might in the presence of such a staggeringly beautiful woman whose almost too casual draping of her splendid figure in a Hare Krishna pale blue robe would have been theatrical in anybody else. It was only the stunning beauty of the woman's face which defied all categorizing and compromise.

The luster of the eyes, the flawless sweep of the features across a well-nigh perfect face, the exquisitely dark, flowing hair, was all one. Catharine Copely was indeed a knockout. Immediately, Philip St George could understand how level-headed, razor-sharp Norma Carlson could have been guiled. Catharine Copely could have sold anyone the Brooklyn Bridge if she put her mind to it. And her powers – who could say what they were?

'Bond?' Catharine Copely murmured in a low, far-away voice.

'That's me,' Peter Bond agreed in a faintly nasal voice, smiling even harder. 'I've wanted to meet you. Show you some of my designs . . . '

The remarkable eyes lifted slightly. Surprise mingled with scorn.

'Really, Mr Bond. Can you possibly be serious? Showing *your* designs to another designer? How could you ever get me interested in anything as foolish as that – or aren't you aware of the fact that I design my own clothes? The entire line, Mr Bond.'

She had paused in her caustically detached contempt of this parvenu because Mr Peter Bond was suddenly behaving very erratically, whoever and whatever he was.

First he had blinked at her in the very middle of her harangue, then lost his smile, came closer, and peered like a man who suddenly finds the names in a telephone directory difficult to read. Then he quickly fumbled in the inner pocket of the rumpled tweed coat, drawing forth a pair of caramel-brown framed spectacles, hastily clamped them over his hawkish nose, and just stood there, staring. Gawking, in fact.

And staring. As if he were facing the unbelievable. A fantasy.

His face metamorphosed. Lost even more color. He swallowed, unable to speak, ultimately shook himself, almost swaying, and then, as some last resort, stiffened his stooped figure and got hold of himself. It was then, and only then, that Catharine Copley made note of two things. Bond had a silver clip which was a pentagram studded with a red stone in the center. In his left lapel another silver emblem shone. An ankh. The tau cross with the loop at the top which always symbolized generation and enduring life. A nation of hippies had usurped the sign.

Catharine Copely's lustrous eyes glittered strangely.

'Mr Bond?' It was almost a feline purr, sliding over the name.

Peter Bond sighed, controlling himself, trying to smile very bravely. 'Sorry – you must think me some kind of character – but, well – '

'Well?' Now fresh interest sounded in her low, sensual voice.

'Look. You'll think I'm crazy. Maybe even think I'm making some sappy kind of pass. I should have seen it right away but I wouldn't put on my glasses. You know how we salesmen are. Always trying to make the best impression right off. Actually, I'm as blind as a bat without them. But, my God, Miss Copely – the resemblance is' – Peter Bond shook his head again, as if unable to credit his eyes, even with the spectacles now perched on the aquiline nose – 'just uncanny, that's all.'

Catharine Copely seemed to forget her sketches and her pique entirely. She folded her slender hands together and leaned forward. The pale blue robe rustled. Her eyes fastened on the ankh and the pentagram tie clip adorning Peter Bond's torso. The red stone might have mesmerized her. The barest traces of a mocking smile touched her serene mouth. When she spoke again, her voice was caressingly soft and gentle.

'Do continue, Mr Bond. Resemblance? To whom?'

Peter Bond's expression was now tragic. Grimacingly sad.

'My dead wife. Amy. You could be twins, I swear.'

'Amy?'

'Uh-huh. Died a year ago next March. I just haven't been the same since. She was everything to me. And, well – I'm sorry you don't want to see the designs. But Frobisher liked them and he said you might want to buy them just to get them out of competition or maybe even incorporate some of the ideas into your own sketches – well, that's show biz.'

He had turned to go, exiting as gracefully as possible following a head-on defeat and an embarrassing incident. Catharine Copely held up a long, graceful arm. 'Wait,' she commanded without raising her voice. The lustrous eyes were stirring in their pools. 'The error in judgment was Frobisher's, not yours. But I would like to hear more about your wife. You say I resemble her a great deal. Can you prove that, Mr Bond?'

Peter Bond, turning, blinked behind his glasses.

'Beg pardon?'

Catharine Copely favored him with a devastating smile.

'Do you have a photo of Amy? You must have one in your wallet, that you carry at all times, since you say you loved her so very much. She was everything to you, I believe you said.'

Peter Bond stiffened as if she had struck him.

48

'No, I don't have her photo on me, Miss Copely. And I'll tell you why. I cry my eyes out every time I think of her. Looking at her picture only makes me feel worse. Satisfied?'

'I'm sorry. I can understand a thing like that. Satan himself never wishes to be reminded of the gates of Heaven from which he was expelled, does he? Paradise seems only more so when we lose it, doesn't it?'

'Lucifer does okay,' Peter Bond mumbled hastily, as if he didn't want her to hear him. 'Don't worry about the Emperor of Darkness. Well, guess I'll go now. Sorry you don't want to see the designs. Frobisher was sure – '

'Frobisher is a fool.' Catharine Copely looked deep into Peter Bond's eyes. The distance was great, she was still sitting behind her desk, but Mr Bond fidgeted under her direct scrutiny. 'Come back tomorrow, Mr Bond. Twelve o'clock. You will take me to lunch. Bring the designs. And a photograph of Amy Bond. I'm certain you at least have a photograph of her, even if you refuse to carry it about on your person. Perhaps we can work something out, yet. All is not lost. You see, I'm rather keenly interested in people who have lost a loved one. Nothing morbid – I like to help them. Meet other people, get out of their grief. Who knows? Frobisher may have done you a favor yet. In spite of his poor judgment, I do not need anyone's designs but my own.'

Bond's sudden, second-chance smile was completely disarming.

'Well – that's real white of you, Miss Copely. And I do have a snapshot of Amy. I have the feeling you still don't believe me, but just wait. You'll swear it's a picture of yourself. As for the designs, I'll bring them, too. I would have brought them today but I didn't want you to rule me out as an eager beaver. Wanted to sound you out first.'

'It is well that you did,' Catharine Copely said. 'I abhor all people who carry their wares with them at all times. It's ill-bred and a throwback to the marketplace. Very well – I'll look but I won't buy.'

Peter Bond nodded but frowned, adjusting his spectacles.

'A salesman has no other choice, has he?'

'They can all go to the Devil!' Catharine Copely snapped in

her only show of heat. Then she smiled, as if realizing what she had said. 'And why shouldn't they, indeed?'

'The Devil needs all the salesmen he can get,' Peter Bond said in a tone heavy with hidden meaning. 'So many people miss his message and don't get the point at all.'

'Strange you should say that, Mr Bond.'

'Why is it strange?'

'Don't you hate God for taking your wife away from you? She must have still been a very young woman.'

'She was and I do hate God. But not for any silly reason like that. Amy was killed on her way to church on a Sunday morning. Can you beat that one? She died because she was *too* religious. Me – I want no part of it anymore. I'm kind of leaning the other way now. . . . ' His voice trailed off.

Catharine Copely closed her eyes suddenly, as if abruptly fatigued.

'Is that why you wear the pentagram tie clip and the ankh?'

'Sort of.' Peter Bond nudged his nose with a hesitant forefinger. 'But that's another story. If you're interested, talk to you about it tomorrow. Over lunch. Deal, Miss Copely?'

'Yes, Mr Bond. Tomorrow we shall talk about many things. Including the photo of Amy. Seeing is believing, I must admit, and it will make me feel much better about you.'

'Okay. If you say so. See you tomorrow.'

'Au revoir,' Catharine Copely murmured as Peter Bond quickly shuffled through the door of her office. She remained where she was for a long time, her dark eyes unblinking, staring at the closed portal as if it held the written answer to a great many things.

When she at last stirred from her statuelike immobility it was to place a call to Karl Frobisher of Fun n' Sun Sportswear to check on the authenticity of Peter Bond. To satisfy her suspicions.

Already alerted by Philip St George, who Karl Frobisher was certain had merely picked a clever method of introducing himself to the most beautiful girl in New York, the wealthy, amiable tycoon of the summerwear crowd easily fielded all of Catharine Copely's seemingly innocuous questions. Philip St George had filled Frobisher in on the entire Peter Bond masquerade of dead wife and dress-designer career, as well as all the details of the phantom Amy's death. As to the Bond appearance, St George had tested it out on Karl Frobisher so he could give a reasonable description, if pressed too strongly by the lady.

Philip St George had once saved Karl Frobisher's life high up on the Andean slopes during a climb of a lofty peak. The rope had snapped but the Satan Sleuth had wrestled life from death by virtue of tensile strength, cool nerves, and quick thinking. Frobisher would do anything for him.

As for the photograph of Amy Bond . . . the nonexistent dead.

The pentagram tie clup which Catharine Copely had seemed to admire had been in reality but another of those technological marvels which Sidney Kite had searched for and found at Philip St George's request. Now it would be no trick at all for Phil, with his own private dark room and photographic equipment, to ink out the Hare Krishna robe, age the print, and make it seem for all the world like a poorly rendered reproduction of a beautiful woman. A tight head shot.

Not even a supposed witch would spot the trick.

It was to be hoped.

The tie-clip-hidden camera, for all its infinitesimal size, was a wonder. A mere touch of a lever atop the circle and – instant picture.

Catharine Copely had never noticed, during all of Peter Bond's fidgeting and near breakdown, that she has having her picture taken.

It was to be hoped.

After all, if Philip St George was ever to find the answer to Norma Carlson's incredibly bizarre cyanide death and link the headless-corpse series of murders to a witch's coven, there was no other way to do it but to join that coven. Get on the inside and see for himself.

He had the feeling that Catharine Copely was sizing him up as a prospective member of that coven which overlooked Central Park.

Somewhere.

Wherever it was.

She obviously needed recruits.

With so many people dying. And leaving the ranks.

He was sure that that was what lunch was going to be all about.

Copely asking Bond to sign in, please.

It was to be hoped.

The Satan Sleuth didn't wish to wait for more people to be killed and decapitated before putting a stop to the wholesale slaughtering spree.

The Devil was having a field day in Manhattan.

Satanism was riding high. Possibly on a witch's broomstick.

Only the Satan Sleuth, who understood the workings of the dark minds of the underlings of the Prince of Darkness, could end the demoniac butchery once and for all.

It was to be hoped.

Book Two:

THE COVEN

'The Black Mass is a valid Satanic ceremony only if one feels the need to perform it. Historically, there is no ritual more closely linked with Satanism than the Black Mass.'

– The Satanic Rituals, by
Anton Szandor La Vey, © 1972

THE INVERTED CROSS

The room, or rather what a Satanist would designate a chamber, was black. Frills, ornamentation, and luxury were not welcome here. Dark, unadorned walls, a vaulted ceiling, and stark, unrelieved bareness was the motif. For this was Sister Sorrow's unholy of unholies, where she performed, exalted in, and executed her own very special and particular version of *Le Messe Noir*, the dreaded Black Mass of ancient and medieval history.

None of Sister Sorrow's brotherhood and sisterhood, which comprised her elect and select coven, had ever dared question any of her modifications and changes in the infamous ritual and ceremony. If anything, even had they been aware of the differences she enforced on them, and the Mass itself, they would only have counted the alterations as but one more darkly illustrious example of her extraordinary powers as a witch. As a power. As a woman of the underworld.

Whatever her motives, reasons, or whims, she was indeed Sister Sorrow. The Daughter of Darkness. Of Hecate. Of Sorcery.

Of course the coven, gathering together on the very next evening, following the banishment and metamorphosis of foolish Brother Carmody, always preferred quitting the bleak, terrifying chamber to continue and conclude the Mass under the stars and moon, out on the dimly-lighted, incense-laden patio bordered by the Bonsai trees and the protective, sheltering bower of shrubs. But in the chamber, at least, the mighty sister had closely hewn to the text and prescription of the ritual book

55

– the Satanic Bible. There was some comfort in that, especially for a true devotee, a sincere disciple of the fallen Lucifer.

As prescribed by the ancient orders and rules, all of the coven, including Sister Sorrow herself, wore hooded black robes. Each took turns serving as priest for the ceremony. The sacred and profane artifacts and implements were always on hand. Whichever woman served as the altar would lie naked on the black platform with her body at right angles to its length, knees parted, legs spread-eagled, a pillow beneath her head, her extended arms holding the black candles which lighted the eerie chamber.

The *Sigil of Baphomet*, the inverted cross, stood forth in bold detail, towering over the ritual in all its meaning and importance, from the back wall of the room. The floor itself, encasing the nude who served as altar, showed the huge, pentagram circle with its five-pointed star of crossing lines and cabalistic symbols. The pentagram was gold-colored.

There was a small organ in one corner from whose deep throat could be plucked the proper somber, liturgical music – Bach, Scarlatti, Franck, and Marchand were most suitable for the mood of the Mass.

Sister Sorrow's coven had come to know and understand the Black Mass in each and every detail and facet of its true meaning.

They all had taken turns playing the celebrant (the priest), the assistant deacon and subdeacon, the nun, the gong striker, the incense holder, the illuminator, and the altar – that most uniquely self-revealing characterization of all. Sister Sorrow had carefully and dutifully explained each of the roles to her flock. Though the membership had suffered many a void recently, as with sisters who failed and brothers who regressed – such as foolish Brother Carmody – she never seemed to tire or the chore. If anything, she gave it all she had; the majestic power of her personality, the fervor of her zeal, the conviction of her belief. When Sister Sorrow expounded the beauties and wonders of *Le Messe Noir*, the coven trembled and yet they *believed*. With every atom of their beings.

As they never had before.

The banker, the theatrical agent, the dietician, the Broadway

producer, the news commentator, the magazine editor, the engineer, the architect, the school principal, the veteran character actor, the commercial artist – they all believed because they had to.

They were under Sister Sorrow's magic power. In her thrall.

It was not difficult for them to recall what had happened to the dress manufacturer and the newspaper-woman, who had fallen out of Sister Sorrow's favor. Charles Carmody was now an insignificant green garden snake and Norma Carlson's strange and unexplained heart seizure – it had made the front pages that morning – well, the coven understood what had happened to all the previous members who had defaulted and lost their privileges. Headless death could be but another form of penalty for failing the laws and orders of Sister Sorrow's coven. They had all learned that lesson immediately. From the very onset of their involvement with the witch. Punishment had been shown to them all. The penalty for backsliding, betrayal, and disbelief was – death. Total obliteration.

It had to be, if the coven was to survive. Live on. Thrive.

For the outside world would stamp them out, crush them and destroy them all if that world but knew of their existence.

The coven tie was as binding as life itself. It was the Alpha and Omega of the soul to become one with Sister Sorrow. By blood oath, ritual of the Black Mass, by obedience. By blind, unswerving loyalty and faith. By accepting Sister Sorrow as Satanically divine.

And so they all did.

And Max Toland became Brother Herod.

Wayne Rilling became Brother Judas.

Bedelia Aaron adopted Sister Lilith as her namesake.

Sloane Gilley was entitled Brother Attila.

Duane Farmer was called Brother Nicholas.

Alice Ainsworth, Sister Circe.

Jason Browne, Brother Mars.

Buck Benson, Brother de Sade.

Wanda Ravelli, Sister Medusa.

Baldwin Simpson, Brother Hitler.

Sophie Smythe, Sister Bertha.

Stupid Brother Carmody had fallen from grace just as he

was about to earn the circle's respect and love as Brother Sawney.

Sister Borgia, Norma Carlson, had been exterminated most fittingly.

And now there were two vacancies in the coven.

A brother and a sister.

The coven had gathered once more at Sister Sorrow's behest to review and discuss the matter although, if the past was any guide, all nominees and aspirants were usually forthcoming from the mind and soul of Sister Sorrow herself. After all, she was empress here.

It was truly amazing, really, but perhaps it was but another demonstration of her magical powers. She seemed able to sense, read into people's spirits, to determine when a person was ready to accept the Devil. To embrace Satanism. She had made mistakes, of course, but she had very quickly rectified them. The coven did not question her choices.

Now, in the stark, candlelit chamber, the *Sigil of Baphomet* glowing with incandescent luminosity behind her tall, hooded figure, Sister Sorrow addressed the assemblage. All stood before her, robed as she was, in a rank-and-file of eleven. Two vacancies were signified by empty places. Subtle musks of incense filled the chamber. The organ was mute. The gong remained unstruck. The altar was absent. The room was still.

The Black Mass would come later.

When there was a new aspirant, or two, to initiate into the circle.

'Tomorrow,' Sister Sorrow intoned in her musical, finely-pitched voice, 'I will interview a man. One who has all the qualifications to join our circle. If he is of the proper credentials and the necessary faith – it has come to me that he has been put in my path by the Master of us all, Satanas! If any of you here know of someone – anyone – speak and declare your sponsorship. It is our duty to enlist acolytes, now that our ranks have thinned, due to the weakness and softness of all mankind.'

No one spoke. Mutely, the figures waited.

Eleven statues might have been facing Sister Sorrow.

The conical hood which concealed most of her extraordinary face and made a shapeless parody of her superb figure with its

58

billowing length seemed to nod. Shadows flickered on the wall behind her, touching the inverted cross. The *Sigil of Baphomet* wavered in the subdued lighting.

'It is well,' Sister Sorrow continued, her voice rising abruptly. 'Now let us designate the execution of the Mass. There must be harmony and persuasion tomorrow night. So that this man, this aspirant, will welcome Satanas with all his heart and soul. You – Brother Nicholas – will serve as the celebrant of the Mass!'

Brother Nicholas, Duane Farmer, bowed. 'I hear, O Sister.'

'You – Sister Circe – will be the incense bearer!'

'I hear, O Sister.' Sister Circe, Alice Ainsworth, genuflected and kept her head down. Her bulky body shivered with expectancy.

Sister Sorrow's robed arm elevated, pointing.

'You – Brother Herod – will strike the gong!'

Max Toland, Brother Herod, murmured: 'I hear and I obey, O Daughter of Beelzebub.'

'Brother Judas, thou will serve as the illuminator!'

'I hear,' Wayne Rilling chanted, hood shaking. *Judas* smiled.

'Sister Medusa – you will be the nun!'

Wanda Ravelli, Sister Medusa, uttered a low moan of ecstasy.

The assemblage, emitting a matching, crooning hum of sound, seemed to sway in unison. A visible termor coursed through the group.

'Deacon and subdeacon,' Sister Sorrow announced, happiness in her voice for the first time, 'shall be Brothers Attila and Hitler!'

Sloane Gilley and Baldwin Simpson raised their arms ceilingward, whispering their acknowledgment of her commands. Her orders.

Sister Sorrow removed her hood.

Her face, seen suddenly like that, was unforgettable.

The weird, flashing eyes, the penetrating beauty, the bold, carmine mouth, the exaltation vibrant in her ivory white face, was plain for all to see. Nowhere was there any woman more striking than she.

'And I, Sister Sorrow' – the magnetic voice soared on a note of wonder, a tone of magnificent ululation, then trembled – 'will be the altar!'

The chamber rocked with movement, whispers of awe. And amazement. A chanting pleasure throbbed among the listening devotees. The Black Mass was never so well served than when Sister Sorrow took the position and role of altar. She always made it a *moment!*

Her incredible beauty, her nude majesty, her faith, her zeal – it would be the richest and greatest of all experiences to attend tomorrow night's execution of the ritual. *Satan be praised!*

The newcomer, whoever he was or might be, was being given a rare honor – the greatest of them all. The ultimate in Dark Mystery.

With Sister Sorrow being the altar for his first Black Mass!

How lucky could some brothers be!

Gleefully, joyously, with voices pitched in matching passion, the coven began to sing. With low, humble, almost debased rhythm that gradually, feverishly rose to a note of mutual adoration and triumph.

The chamber quivered with song. Demoniacally rendered music.

A song of Sister Sorrow's own devising. A pæan, a hymn. Of love.

And personal worship to the Devil below. Dark Lord Lucifer.

'. . . *I reign on coals where live damned souls*
Among my hordes of shoveling wards.
But though I am feared, my soul is seared
Because of Him who forsakes Me. . . . '

As the pitched voices increased in volume, Sister Sorrow raised her arms, facing the *Sigil of Baphomet*. She loudly and whimperingly called out Bauldelaire's passionate appeal to satanism:

'*Satan, have pity on me in my deep distress!*'

The chamber sent back the echoes of the fervid plea, the words stealing about the dark walls of the room, profanely.

The coven moaned its agreement.

And its unswerving fealty.

All in chorus.

All together.

60

As one.

Their souls, minds, and hearts were locked in with Sister Sorrow's, *with* the Devil and *against* the world at large.

The world which refused to understand and acknowledge the Prince of Darkness, the Emperor of Everything.

More fools, they!

Little did they know the glories they were missing.

What paradise they eschewed.

Philip St George had had a very busy day following his strange interview with Catharine Copely, the stunning woman who was the head of Chic, Incorporated. Seldom had he planned a campaign for action with more careful preparatory provisions.

He had taxied directly to the Americana Hotel, registered as Peter Bond, listing his home address as Allerton, Pennsylvania, and his occupation as fashion designer. On his person he carried the sufficient IDs to back up the characterization. Diners Club credit card, Master Charge plate, driver's license – all of which were authentically in order and serviceable. The St George millions were sufficient to provide a false Peter Bond with legitimate records and a physical history which could be checked on at any time. By anyone, for any reason.

Philip St George had covered his tracks extremely well. There was even a set of Peter Bond's fingerprints on file with the Federal Bureau of Investigation in Washington, DC. Of course, they weren't Philip St George's ten prints; the technological experts which St George money could buy had managed the miracle of installing a bogus set of identification plates in the Washington files. Including a fake photo, Army record, school history, and complete family background. So Peter Bond could be investigated and validated by anyone who was interested. Philip St George would only retain the helpful masquerade until the fatal day when any connection was established between such a phantom and the man who had won the admiration of the world with his dazzling feats of exploration and sensational capture of the four ghouls in human form who had butchered Dorothea Daley St George in 1973.

In other words, Phil had given Catharine Copely a trail to follow, a man to corroborate if she had any suspicions at all concerning the oddball who had wandered into her office that day, with his queer reactions.

Till then, he would play the game as Mr Peter Bond, in town temporarily to sell some designs to a lady genius in the fashion field, with transient residence at one of New York's best hotels. With Karl Frobisher assisting in the expanded lie, there was little to worry about.

Unless, of course, Catharine Copely was that sort of witch with genuine powers who could look right through a man and see the great lie.

Philip St George didn't think so.

He hadn't believed in devils or werewolves; he wasn't about to accept witches either. What he did accept and had to fight was the influence such people had over the ignorant, the fearful, and the superstitious. Which, in effect, was his *raison d'être*.

The Satan Sleuth's motive for a crusade was the debunking of bunko. The squashing and exposing of any kind of occult racket.

Once established at the hotel, he made certain that the bell-hops, the porters, and the pretty registration clerk would have little difficulty remembering Peter Bond. This was rather nicely accomplished by a vicious, headlong sprawl across a piled stack of waiting luggage lying in his path on the way to the registration desk. Anyone else might have been badly injured, but a trained, physically coordinated Philip St George managed the harmless accident with an expertise worthy of the best of Hollywood stuntmen. The 'accident' would not go unrecalled.

Then he checked into a fine room on the fifth floor, ordered lunch, placed a call to a well-known dress manufacturer in Allerton, knowing fully well that the man was on vacation in Bermuda, and rested on his activities. After lunch he quit the hotel and cabbed to his own apartment in the heart of the city. A lofty penthouse bachelor affair high above the stone canyons of Manhattan. There was no need to remove the Peter Bond disguise. Not until he was alone in an unattended elevator riding up to the top floor. And even then there was little danger of discovery. The building was accustomed to Philip St George having all sorts of visitors. He had his privacy, as any young millionaire can easily insist on and pay for. No one would ever notice a nondescript character such as Mr Peter Bond.

Philip St George had a lot more to do during the remainder of that important day. Apart from being on hand to receive any calls from Sidney Kite and that all-too-crucial material in the headless corpses case, there was the matter of the film. Catharine Copely's picture.

The red stone in the center of the pentagram tie clip was the first consideration in the master plan to invade the coven.

It could be a far more dangerous game than the old 'Candid Camera' TV show had ever been. In more ways than one.

The amazing little hidden camera had to produce a photograph which would both intrigue and deceive Miss Catharine Copely.

Or the whole play would be lost. Gone forever.

Unsuccessful deception could muff the only chance to penetrate the heart of the mysterious coven and the opportunity to flush out the whole bunch of Satanists. A chance that might not come again.

That was an opportunity the Satan Sleuth did not care to miss.

Not in this lifetime if he had anything to say about it.

Fortunately, he did.

THE HEADLESS CORPSES

Toward late evening that eventful day, while Sister Sorrow was outlining the order of the Inner Circle for her own version of the Black Mass, and Sidney Kite and Miss Irene Walters were busily accumulating a hefty pile of research material required by Mr Kite's number-one client, and the Police Lab. down at Headquarters was trying everything in its forensic powers to determine how Elvira Davis, the fifth victim, had gotten that way – the wheel took another spin. Perhaps the strangest rotation of all. One wild whirl.

The Manhattan branch of the Federal Bureau of Investigation received an anonymous, incredible phone call, very late in the afternoon.

A taping device immediately began to reel in the conversation. Later, Fling was very grateful for standard operating procedure. He had instinctively pegged the brash, oddly familiar male voice as worthy of coming under the heading of Crank-Call-Informant Potential.

For, as matters developed, the anonymous caller was a fund of vital information. Terry Lenning would have loved him. But Lenning was away from his desk at the time, and could only hear the conversation secondhand a day later when Fling ran it off for him, asking his opinion. Unless there was a leak somewhere at the top, the caller could not know what he knew. Unless he was just a nut, after all.

Or the murderer. Murderers could be telephone nuts, too.

In any case, the dialogue was an ear opener. And eye opener, besides, when you played it a few times. Read weirdo, screw-loose stuff. But good.

The official transcript, for reading, as well as the spool of tape, was about seven minutes long, completely intelligible, and thoroughly substantiated most of what Terry Lenning had already officially guessed.

Somehow, that was the most nettlesome item about the call. Like hearing a playback of someone else's private thoughts.

Arthur Fling was astounded and made not a little nervous by the dialogue he shared with a voice coming seemingly from out of the blue:

Chief: 'Who is this calling, please?'

Voice: 'Never mind that. I want to talk to you about the headless bodies you've been finding all over the city.'

Chief: 'Sorry, that's New York Police Department business. Out of our jurisdiction.'

Voice: 'Wrong. I happen to know your bureau now looks into anything of an occult nature where terrorism and possible underground societies are involved. Don't stall around about it, Mr Fling. I could give you the law, the bill number, and the name of the senator who introduced that fine piece of legislation only last year.'

Chief: 'Oh, well, it's still a local police matter. Why not tell me who you are? I am willing to listen to you.'

Voice: 'Stop it. You won't be able to trace this call. Take my word for that. Nor will your voice-print machine place how I really sound. As you may or may not have noticed, I'm imitating Gable and then I'm going to switch to Bogart, then some Cooper and a little Cagney before I hang up. Forget about the police department. They haven't the imagination or skill to handle this matter.'

Chief: 'I noticed. Not bad. Well – all right then. Since you seem to have finessed me from the start, what is it you want to say? This is a busy office, my friend.'

Voice: 'That's better. I'll talk. You listen. And don't interrupt until I ask a question. Firstly, I have no idea what the official headquarters' line is in these deaths, nor do I know what you FBI people are up to, but the occult and Satanism are my business. So I wish to give you the benefits of my considerable experience. Now, have you got all that and are you ready to listen without further interruption?'

Chief: 'Yes.'

Voice: 'Good.' (The Gable sound suddenly replaced by Bogart here.) 'Your five women were murdered by a Devil-worship cult. Given the abandoned nature of the scenes of the crimes, the nudity and mutilation, as well as the crucifixion poses and decapitation – there is no other possible answer. Removing the victim's head is an official act of a Satan cult. In the Black Mass, which to Devil worshippers is as High Sunday Holy Mass is to a Catholic, a malefactor or unbeliever or anyone who has fallen out of the good graces of the cult – *is always beheaded.* So look no further. This is not the murder binge of a single person, a psychopath with a blood mission, or any such mass murderer solo performance. This has to be a group effort. I'm sure you gathered that yourself, from the MO of the murders themselves. Physically, they would be impossible for one person to carry out. Now – do you agree with that fact?'

Chief: 'You seem pretty sure on that point.'

Voice: 'Don't be an idiot. Have you ever tried to decapitate a person without having to tie him down or hold him down? You found no evidence of that according to all I've read about the cases.'

Chief: 'All right. Murder by group, you say. A Satan cult. You still say. We see it that way too. Satisfied? So what?'

Voice: 'So this. Those five women belonged to one of the most dangerous cults in this city. They were killed because they lost faith in the Devil – or because they had something the leader or leaders of this cult wanted. Now, I intend to join that cult, discover who its leader or leaders are, and then I will tell you – but first I must find out where the group holds its meetings. I also intend to learn why these women were killed. Five females without kith or kin left in this world. All wealthy – there must be something in that. There has to be a profit motive, you see. I don't believe in the Devil and it's been my experience that those who do usually have some simpler goal in mind – like making other people believe in the Devil – and making that belief pay off some-

66

how. Like in dollars and cents. A great deal of dollars and cents, Mr Fling.'

Chief: 'Your Cooper is better than your Bogart, but you shouldn't change actors in midstream like that. It distracted me. Very well. You're a public-minded citizen who knows all about these things and you're going to play hero and stick your neck out and then call me at the last second to come running in like the Marines to bail you out. Be reasonable, whoever you are. Come on in, talk to us, tell us all you know. How am I going to know it's you when you do make that call? How do I know you're not the murderer yourself and you're just playing psycho games with me to get your kicks? It's happened too many times before – you characters like to call up and brag about your kills.'

Voice: (Cooper imitation replaced by Cagney.) 'Oh, I'll call you again. Never fear. Just don't dismiss me as a screwball. You'd be making a terrible mistake. When I do phone you again, I shall give you a password. Pass*poem*, really. It shouldn't be too difficult for you to remember.'

Chief: 'What might this passpoem be?'

Voice: *'I am a man. I am the truth. You can call me – the Satan Sleuth.'*

Chief: 'Hold on now – give me that again.'

There was no answer. The cartridge picked up molecules of empty air. The odd phone call, and the transcript, ended there.

When he hung up Arthur Fling tried manfully not to swear, succeeded, but then yielded to pique, snapping a Number 2 Mongol pencil in half. His subsequent sigh of helplessness exploded in the cool atmosphere of the orderly and tidy office. A blast of vexation.

If he didn't know better, he would have sworn that Terry Lenning had just called him up, disguising his usual voice four times, just to reinforce his own pet theory once more. About the wave of headless corpses dotting the Manhattan landscape. But he dismissed that hare-brained thought as soon as it occurred to him. Terry Lenning's mental level of dignity and common sense made such a conclusion unthinkable. No, the call was on the level. It had to be – crank or not, murderer or not. Clever, whoever he was, masking his voice like that. Imitating movie stars would play havoc with the voice-prints.

The Satan Sleuth?

Now what the hell was so naggingly familiar about that name?

Chief Agent Arthur Fling was dead certain he had heard that peculiar title once before. Heard it or read it or –

For the life of him, alone in his orderly office, he could not remember where. Or when. *The Satan Sleuth . . .*

There were just too many things to think of, take care of, and dispose of when you were the head of the Manhattan branch of the Federal Bureau of Investigation. Things piled up, cases backed up.

But he could find out. Ask somebody. Look up the name in the prodigious filing system at his disposal. The whole mammoth machine known as IBM. All the way back to Washington, DC, if need be.

And Chief Agent Arthur Fling would do exactly that.

When he had the time. And the opportunity. And the notion.

Meanwhile there was much to ponder over.

Thanks to one mysterious phone call from a voice from out of the void. A call in a city of over ten million people. All with phones!

The Satan Sleuth.

Sweet Mother of J. Edgar Hoover!

That sounded like something right out of a comic strip.

Batman or *Captain Marvel* would have made just as much sense.

Still, you had to take calls like that seriously. All the same. The voice, whoever he was, had made an awful lot of logic out of five seemingly senseless, bloody murders. A cult added up.

Arthur Fling had not as yet made any connection between the weird slaughter spree and Norma Carlson's mysterious demise in a Manhattan restaurant while she was out dining with a male escort.

Nobody else had, either.

Nobody official, that is.

Unfortunately.

In the humming, monster concrete monolith and beehive known to the world as Grand Central Station, Philip St George emerged from the third telephone booth in a stately bank of five such boxes on the underground level of that mammoth landmark. He moved casually.

He still wore the face and clothes of Peter Bond.

And he was smiling, almost tolerantly, as if bemused by the usual, familiar spectacle of his fellow human beings scuttling back and forth on errands of departure and arrival. Actually, his pleasure lay elsewhere. Mostly in his own churning, methodical mind. Things went well.

The call to Fling of the FBI was completed. All the wheels and dials were now set for turning. It only remained for Sidney Kite to come up with some important data about the five female victims; for the police to determine exactly how the cyanide had been administered to Norma Carlson so impossibly and fatally; and for Peter Bond to successfully infiltrate the strange Satan cult which seemed to be governed by the striking woman known as Catharine Copely. Only that.

The photograph produced from the tie-clip camera had been exactly right. Perfect for touching up, cropping, and continuing the deception that Peter Bond's dead wife, Amy, closely resembled Catharine Copely. Enough like her to be a twin sister. Miss Copely would never be able to spot the photo as a picture of herself. Not after Philip St George's magic touch with the negative in the dark room of his home.

It would take more than a witch to call the camera a liar.

The day had been eventful, details had been accomplished; there was no reason not to expect a rapid fruition of the plan to thwart a Satan cult and drive it into the open, where exposure would ruin it for all time. Creatures who thrived in darkness couldn't seem to survive in the glare of daylight. The most fundamental laws of nature never applied more aptly than when visited upon shadowy objects. Sun beats moon.

The same rule of thumb would fit a witch and her coven.

At least the Satan Sleuth thought so that crucial day in the baffling, murderous case which was horrifying a city of frantic, worried multitudes who thought they had seen and heard everything.

What Philip St George didn't know was that somebody had been following him most of that day. From the moment he had left Chic, Incorporated he had been dogged at all times by a most persistent shadow. Another of Catharine Copely's shadows. The darkest one.

A tall, somber-faced man in a Henry Higgins hat, blue blazer under a floppy trenchcoat, and half boots tucked over very tight gray slacks. The man had been with Peter Bond all the way. Waiting as he registered at the Americana hotel, following him to the midtown skyscraper apartment building, and again lingering until he came out once more to walk to Grand Central to make a phone call late in the day. Philip St George had not picked up his shadow once, for the man possessed nearly supernatural ability to remain unseen and undetected.

Had Norma Carlson lived long enough to tell Philip St George more of what she knew, she might have mentioned some of the outstanding physical features of Duane Farmer.

For one thing, his craggy face was funereal – the cliché cartoon representation of all undertakers and morticians everywhere.

For another, he had a nose that was not only formidably large but bore an exceedingly close likeness to the pistol grip of an old Western shooting iron. It was just as weathered, just as veined.

But more remakable than all of that, and far more memorable to anyone who had ever spent as little as five minutes in his presence, was a pair of compelling, deep-set eyes whose woeful sadness and dire gravity could easily have foretold the approach of the end of the world. Armageddon, Doomsday – complete annihilation of the human race, in short. Total destruction all down the line.

This then was the individual who had introduced Norma Carlson to the woman known as Catharine Copely. ' . . . *as ugly as sin, Phil.*'

This also was the man who served as her second-in-command, her closest confidante in the Inner Circle. ' . . . *I listened to what he had to say.*' Duane Farmer – Brother Nicholas – who more often than not filled the role of celebrant of the Black Mass. The priest himself.

Philip St George had underestimated Catharine Copely.

Badly – almost recklessly.

When he met the woman again the next day, in the glossy environs of Chic, Incorporated, for their getting-to-know-you luncheon date, he was to be at a terrible disadvantage. Despite his precautions.

Catharine Copely already knew far more about Peter Bond than he might ever know about her.

She knew enough about the man called Peter Bond to know that he was her enemy, whatever his game really was.

That in itself was enough to kill people.

To kill anybody in the whole, wide world of disbelief.

And unbelievers.

It was her sworn duty to scourge the skeptic, destroy the infidel, expunge any mortal who dared defy the Bible of Satan. Especially if that unfortunate chose to make some kind of war against the Prince of Darkness and those who worshipped him.

Catharine Copely, also known as Sister Sorrow, was just the witch for that assignment. It was right up her nightmare alley.

She had a coven. A loyal, blood-thirsty, blindly obedient band.

And she had the tools and means and wherewithal to make come to pass anything she wished and desired. Witchcraft could make it so.

Philip St George was about to learn that lesson the hard way.

Unforgettably.

To his everlasting regret.

He was about to descend into the hell he had never believed in. The one he was certain existed only in people's minds.

God help him.

Duane Farmer followed him once more and did not relinquish the pursuit until Peter Bond had again disappeared into the lofty skyscraper eyrie in mid-Manhattan. The thousand and one lights of the city glowed like fireflies in the misty, windswept winter night.

They could have been danger signals, warnings flashing in the night, had Philip St George not been blind to their message.

But he was only human, after all, not a demon.

Unlike Catharine Copely, the Sister Sorrow who was spiritual leader of the hellish coven which held its ritual meetings in the sky-high penthouse overlooking sleeping Central Park.

The coven that dripped blood.

LUCIFER'S LADY

Long before the first touches of dawn fingered the eastern sky showing through the bedroom windows, Philip St George had been awake. Reading far into the night. Boning up on facts half known, possibly forgotten. All about his pajama-clad form, softcover books and hardcover editions lay in scattered profusion on the large French provincial bed. He had not returned to the Americana, feeling no need to carry the Bond impersonation that far. Instead, he had holed up in his penthouse apartment and busied himself, making ready for the crucial return engagement with Miss Catharine Copely.

He had had absolutely no inkling or suspicion of the ghostly shadow which had kept pace with him most of the day before.

Duane Farmer might have been transparent – or invisible.

There was much to do if Peter Bond was to successfully continue deceiving Catharine Copely. The trick photograph was all prepared, tucked in a five-by-eight manila envelope, to show to the beautiful lady.

But there had to be more, didn't there? Peter Bond had mentioned a set of designs for showing and possible sale, if the lady liked them.

So Philip St George, for whom all skills had become a matter of course in a lifetime which had known the best schools, all the advantages, coupled with his own incredible physical dexterity and genius-level IQ, had set himself to the task. Artistically.

In the mere space of two hours he had rendered a dozen preliminary sketches which would more than satisfy a Catharine Copely. In Conti crayon, with a flowing beauty, and utilizing the best features of St Laurent and Gernreich, borrowing freely, Philip St George had produced a fine collection of feminine wear. Once the artwork was completed, stacked, and tied in a fourteen-by-seventeen folio of the usual type, he was able to take a luxurious shower and make himself a small meal of cold chicken and tossed salad, accompanied by a splendid Chablis. He brewed a pot of coffee in the ultramodern kitchen, spent the time while the pot was perking doing some Yoga exercises designed to get the maximum relaxation for all of his muscular and mental processes, and then retired to the bedroom. There, with the twinkling stars still showing in the brilliantly midnight blue sky, he entered the bed, propped up the huge pillows, and set himself for a very long session with a pile of books within easy reach on the end table. The room would have delighted any woman, for it was so eminently masculine, with deep, woodsy furnishings, a wall-to-wall sculpted rug of rich carmine shades, and a single painting. This, a large-sized portrait in glowing colors, faced the bed.

The loveliness and indelible fresh beauty had been captured on canvas by the renowned artist Merrick. For all time, for anyone to see.

Forever. So that it might never disappear or go away.

Philip St George had never taken down the painting, not even after the horror up at Lake Placid. A heartbreak ago.

He never looked at the portrait anymore, though. Not head-on.

It wasn't a conscious act on his part, nor an evasion. That was simply the way it was and sometimes, as he rose from the bed or turned in the direction of the painting, he was able to see it without a flicker of any kind showing on his handsome face. Even when he was alone, his powerful self-control never slipped. He could never cry again. He had not cried when first he'd viewed Dorothea's outraged remains on the floor of the big house up there in the country – he would not cry at all – for that was the very moment, the single instant, when Philip St George, devil-may-care adventurer-playboy, had transformed into someone else.* A new kind of avenger. A rare

*The Satan Sleuth No. 1 by Michael Avallone, Warner Paperback Library, © 1974.

breed of samaritan. A man for Good, not Evil.

The Satan Sleuth. A name just beginning to assume the iron of legend.

And it had been the proper choice of role for him, as the future showed that such a man was needed. The world was sick and being made a whole lot sicker by Satanic forces. The Devil was abroad in the land. Perhaps stronger, darker, and more insidious than ever before.

Philip St George no longer regretted his past. He couldn't.

The present needed him – as did the future. It might always need him.

Now that Satanism had shown its ugly, blood-red face again.

Five women were dead. Brutally murdered. Defiled by cultism.

A sixth, Norma Carlson, had also died. Even more weirdly.

For probably the same reason, the identical motive. Time would tell.

So Philip St George was doing battle once more. In fact, he was arming himself, as usual, by doing his homework. The necessary kind.

Reading everything that was printed on the subject of the mystic arts. It would be wise to be up on absolutely everything there was to be up on when he met Catharine Copely again. To play safe.

Those sinister arts, especially Satanism, witchcraft, and possession, were perhaps never more frightening, or seemingly more tangible and vivid, than in the night hours. Acts, thoughts, and visions scoffed at in broad daylight can take on an incredibly realistic value when the bright sun goes down. The human mind at best is a fragile stronghold, subject to all sorts of assault when the convolutional, labyrinthine thinking process has the upper hand. It is altogether a very natural outcome of cause and effect in the condition known as life. So it can be that the scientist who ridicules mysticism under the microscope in his laboratory may be the very same man who will think he can get seven years of bad luck by breaking a mirror when he gets home that night. Or refuse to put a hat on the bed or open an umbrella in the house. Superstition is the very first break in the barrier the mind sets up against impossible creatures like demons, phantoms, ghosts, and devils. Things that go bump in the night.

Not so Philip St George, however.

Since the day he had undertaken his bizarre crusade, turned to solid, unflinching vengeance by the gruesome spectacle of his wife's mutilated, headless corpse strewn all about the floor, there had never been any room in his heart, mind, or soul for the mythos of the supernatural. He knew it did not exist just as he knew there was no large, unknown planet in the vicinity of the moon.

To feel otherwise would have left him vulnerable. Afraid.

The experiences of a lifetime had convinced him that all the mystical, phantasmagorical, allegorical creatures of history were nothing more than the fictional creations of men. Baffled, confused men who gave body and form to all the mysteries, wonders, and misfortunes of the universe by assigning those mind-produced images names and titles: Lucifer, Satan, Old Nick, banshees, werewolves, monsters, ghouls, poltergeists, spirits, vampires, invisible men, Abominable Snowman, ghosts, leprechauns, fairies, elves – the whole lot was nothing more than mankind's cop-out for its lack of answers or explanations of natural phenomena and marvel. Parapsychology, ESP, mind-reading, mental telepathy, kinetics – more fun and games of the same sort, even granting that the purpose and motivation might be loftier.

The world had hypnotized itself more than two thousand years ago and the mass hypnosis had never quite left the human race.

And because of that single truth, that unalterable, unchanging paralysis of normal logic, there were people who could use man's fear against him. Sorcerers, palm readers, gypsy fortune-tellers, phrenologists, handwriting experts, astrologers, witches, warlocks, magicians, wise old men, seers, prophets, mentalists – gamesters of all kinds, every variety, ready to turn fear into profit and money. And lifestyle.

Philip St George would never believe in the mystic arts.

But he would read about them. To defeat the enemy it is always necessary to know exactly all there is to know about that enemy. So the shelves of the apartment high in the night sky was always stock-piled with each and every conceivable title on the subject. The Satan Sleuth owned them all.

In the beginning he had designated Sidney Kite to acquire many a book for him, but as time went on he found it simple enough to replenish his occult library as the need arose. Or the interest.

Recent additions to his shelves were *Witches and Werewolves, The Druid Religion, In Search of Dracula*, and *Lovecraft Mythology*, joining such cornerstone books as *During Sleep, Possession, The Satanic Mass, The Book of the Dead, The Satanic Rituals, Hail, Lucifer!, The Magus: A Complete System of Occult Philosophy* and *Gypsy Demons and Divinities: The Magic and Religion of the Gypsies*.

There was very little he had not read on every aspect of the subject. One which embraced countless variations and substitutions for what was, in essence, the eternal search for a new God – and gods.

Perhaps for the same God, in so many different languages.

In any case, reading all about the dark field whose lists he had entered at his own choice and risk was mandatory.

It was while he was deeply immersed in the heavy subject matter during the small lonely hours of the night that he heard the next turn of the frightening screw. In all its terror and mystery.

A sudden burst of strange sound, unusually loud in the peaceful stillness of the bedroom, reached Philip St George's keen hearing.

There was a fluttering, clattering noise, mingled with a crying, almost whimpering screech, coming from the wide window at his right. Instantly his startled glance swung in that direction.

As seasoned as he was to danger and the unexpected, he could not repress a grimace of surprise. Surprise and bewilderment. Plus awe.

Flapping wildly at the clear glass of the window, seeking entrance one way or another, was an enormous, black, darting shadow.

A shadow that should not have been there. No way!

Against the stretching darkness of the midnight sky was

something unmistakable, for nothing else on earth looks quite like it.

Or ever will, come what may. It is uniquely one of a kind.

The jagged-winged, irregular silhouette of a bat in flight.

Black, violent, and vicious. Calling down echoes of nightmare.

Two gleaming eyes, flaming like burning coals, lit up the darkness. The awful screeching sound was the creature's whimpering voice.

Philip St George's own eyes narrowed and his breath caught sharply. A bat in Manhattan, high above the stone street, battering about as if the luxury apartment were a belfry!

The Satan Sleuth moved swiftly, with fluid resolution and trained reflexes. The commotion at the glass was reaching a point of frenzy.

He swept open the grained drawer of the night table, found the sleek, long-barreled air pistol always in position there, and sprang to the window. The bat's hidden face, with only fiercely gleaming eyes visible, was now inches from his own. Cautiously, tensed, air gun poised, Philip St George unlatched the casement and stepped back quickly, pulling open the left window as he retreated. He leveled the air pistol at arm's length, both hands clasped about its smooth butt, in the approved firing-range style. He wanted no more than one shot, as noiseless as the weapon was. The bat might or might not be rabid, but it was nothing to have flying around the bedroom in a frenzy of flapping wings. Though somehow he could not erase the sensation that this was something much more than a mere bat. He could not have said why. But the fluttering creature was kicking up a frightening, noisy storm on its side of the glass. With demoniac persistence.

Philip St George was not prepared for what happened next.

With the window wide open, the night air seeping in with chilling, windy cold, the bat did not react as it should have. It went haywire.

With lightning speed.

The patio beyond the windows ended with a low stone wall which ran at an oblique angle before it joined the corner connected to the adjoining building. Privacy was obtained by a ten-foot-high ivy-covered terrace. The bat, with no window barring its way, rocketed upward, emitted one bleat of sound, and vanished from sight. Philip St George leaped in pursuit. But he was too late to chance a shot at the weird apparition which had disturbed his reading.

He was only in time to witness an aerodynamically impossible performance. For anything that must obey the laws of gravity and physics. No matter what it was or could be. Outside of the Devil!

The black bat soared upward and leaped from view, as if plucked like a kite on invisible strings. *And it was gone!* Leaving only the cold night air, the twinkling stars, a half moon staring down from the mighty heavens and – a sorely perplexed Philip St George.

He lowered the air pistol, frowning.

It was useless to look any further. The thing was gone. For good, perhaps. Back to wherever it had come from. He had no doubt that it was a bat he had just encountered. But what *kind of bat* exactly – he wasn't too sure. Witchcraft hung in the atmosphere, still. As cold as the graveyard, charged with fire and brimstone. There was a deathly, clammy aura to the midnight air. He stared up at the silvery moon, briefly. Unnerved.

Then, closing his pajama top about him, he strode into the bedroom and locked the casement windows. The formidable frown had not left his classically handsome face. He was puzzled.

A bat.

A genuine bat.

Acting like something out of a science-fiction movie. Or Hell itself.

Restraining an involuntary shudder, he replaced the special-make air pistol in its drawer-bed and clicked off the hooded lamp on the end table. There was no more reading to be done that night. He needed a few hours' sleep, at least.

And there was so much thinking to do. About a lot of things.

Philip St George lay in bed a long time, listening to the winds whistle eerily against the sides of the building, aware of every rustle and whisper of sound. Smoking had never been one of his acquired habits. He had never needed tobacco or alcohol.

He had needed only excitement, adventure, and love. And in that order, too. Which was why he had always blamed himself for Dorothea's death. And found it so easy and right to go on his own peculiar crusade. Perhaps if he had loved Dorothea more she might never have been alone in that big house up there in the desolate woods. *Perhaps, if, maybe, only* – words that could torture a man. But Philip St George did not torment himself anymore.

He tormented those who would torment others.

Satanists.

Witches.

Warlocks.

Voodoo doctors.

Hexists.

And just about anybody who used the Devil for an excuse.

But –

There still was the enigma of the bat.

He was still thinking about that, rejecting one solution and then another, before sleep at last closed his eyes. His dynamo had run down.

The bedroom was like a silent tomb in the quiet fastness of the night.

Sister Sorrow stood before the ceiling-to-floor mirror in the stygian gloom of her bedroom. She was praying.

There was only the flickering yet steadily blazing illumination offered by two tall black candles thrusting upward from wrought-iron stands to either side of the gigantic looking-glass. The light was hellish.

There were two Sister Sorrows in that chamber. Paired perfection.

The one in the mirror; the one standing before it. With naked arms uplifted, eyes closed, superbly shaped body glowing palely in diaphanous chiffon of midnight black. The ivory contrast of Sister Sorrow's flesh, tinged with the roseate flame of the twin tapers, was perhaps a vision, a fantasy, that few mortals would ever be given the privilege to see. The Sister Sorrow who walked and writhed nude in the heart of the Inner Circle at the Black Mass was not this one. Not this woman, alone in her bedroom, luminous eyes closed, long black hair trailing, impeccable figure at straddle-legged exhortation, red, red lips moving in an almost silent prayer of invocation to the Lord God Lucifer. Not even Brother Nicholas had ever been granted the pride of seeing Sister Sorrow at her most fervent, Devil-worshipping best. The Divine Sister was surely a goddess when she was like this. An entity, all-powerful. A One.

With only Satanas above her in the heirarchy of the under-world.

Catharine Copely had disappeared truly.

This was Sister Sorrow's incarnate.

Doubled, magnified – no, trebled in the mirror on the wall.

The reflected image was unearthly in its ethereal yet vivid beauty. For these were not pastels or water colors, these were the deep-dyed scarlets, crimsons, and vermilions of throbbing vitality and unholy energy. Sister Sorrow's supremely regal form seemed to evanesce in the crystal clarity of the mirror. Her flesh moved, the limbs convulsed, her rounded bosom strained at the chiffon. The muscles in her hauntingly unfor-gettable face quivered. When she opened her eyes at last, staring at her own reflection, it was as if rockets flared and Roman candles ignited the gloom of the boudoir.

Her long, lustrously black hair, the pools of her amazing orbs, the startling whiteness of her skin, touched with the flame of the tall tapers, was unmortal, fantastically unreal.

Anyone seeing her like this would have believed she was truly a witch. Anyone. Perhaps even Philip St George.

No one could be as beautiful as Sister Sorrow was – *but she was.*

No one could look into the secrets of life and death – but Sister Sorrow could. No one could have such incredible eyes – but Sister Sorrow did. If there could be a daughter of the Devil, Sister Sorrow was certainly that woman. Her belief in herself and her own powers transcended religion. Worship. Evil. She was indeed a lady for Lucifer.

Now she brought her long arms down to her curved hips and brought her legs together. She continued to peer deep into the face of the purely hypnotic creature staring back at her from the depths of the looking glass. Her red lips moved. Her eyes glinted in the candlelight.

'. . . *in nomine Magni Dei Nostri Satanas. Introibo ad altare Domini Inferi* . . .'

The Latin words barely whispered aloud nonetheless seemed to fill the darkened room. Sister Sorrow's voice held a richness of tonality and fervor that could not be denied by mere acoustics.

Sister Sorrow daily repeated the opening ceremony of *Le Messe Noir*. This in itself was a ritual with her – repeating the sacred vows of devotion and faith before retiring for the evening. These were the inspiring words, when all the coven were assembled and the gong had struck and the altar was in readiness. The celebrant, the deacon and the subdeacon approaching the altar, bowing very low . . . and then the celebrant fulfilling his priestly office by saying:

'*Before the mighty and ineffable Prince of Darkness, and in the presence of all the dread demons of the Pit, and this assembled company, I acknowledge and confess my past error.*'

Latin and/or English, the words were stirring, sacredly divine!

'*Renouncing all past allegiances, I proclaim that Satan-Lucifer rules the earth, and I ratify and renew my promise to recognize and honor Him in all things, without reservation, desiring in return His manifold assistance in the successful completion of my endeavors and the fulfillment of my desires.*

'*I call upon you, Brother, to bear witness and do likewise. . . .*'

The coven would repeat the wondrous words in homage and fealty.

Only the altar remained mute and unmoving. Deathly still.

In full sight of the *Sigil of Baphomet*, the inverted cross.

Sister Sorrow's breathing quickened, her breasts rose and fell. She undulated, her tall body twisting in an abandoned dance of passionate origin. Satanism was meat and drink to her. It sustained her.

She began to croon in a low monotone. She closed her eyes again, as if in shutting them she could see all that she wished to see. No one would have recognized the flat, atonal sounds emanating from the sensual lips. Except perhaps another witch.

It was not a song anyone had heard in four centuries. Not since the dark, dread days when Cotton Mather had had his way and burned people at the stake in old Salem. No one in Sister Sorrow's coven had ever heard her sing like this. Not like this, not this particular song.

' . . . aga . . . tor . . . romni . . . du . . . oh-gah . . . oh-gah . . . roo . . .'

Tonelessly she crooned on. In self-inflicted thrall, holding her own shapely arms by the elbows. Swaying before the mirror, undulating in pagan, Satanic release to her master. Her beliefs.

It was then that something stirred in the darkened room.

Sister Sorrow still had her eyes closed so she was not aware of the subtle change in the room. At least not immediately.

Not until the ebony ball of curved fur walked stealthily into view, caught in one lower corner of the floor-to-ceiling mirror.

A large, very handsome cat, almost Satanically feline in its superb coal-black pelt, burning green eyes, and stalking majesty. The animal murmured, the barest breath of a sound, and nudged its soft, caressing body against the naked ankles of Sister Sorrow's legs.

She opened her eyes, nodding as if to herself, whispering, 'You came. You know . . . you know everything there is to know, don't you? . . . Sweet, precious familiar of mine . . .'

The cat purred, arching its tufted back, green eyes slitting.

Sister Sorrow reached down slowly, cupped him in her slender hands, and raised him gently aloft. He did not protest or fight back. He knew what she would do. She did it. Perching him serenely on one advantageous shoulder, Sister Sorrow stared at the cat in the mirror. He stared back at her and himself, green eyes inscrutable.

As unfathomable as a stone idol in a museum.

Sister Sorrow did not seem to mind the interruption of her witch's rune. There would always be time for singing. For runing.

The black cat was her most beloved of all fleshly things on this earth. It had been with her since the very beginning.

'What is it, my Poe? What troubles you?'

The cat did not answer, throbbing contentedly somehow.

83

'What see you in the mirror? Do you recognize a false image? Do you see that the infidel who has come into our midst is not what he pretends to be? Can you also know, as I know, that the man Bond wears a false face? That he means all of us harm?'

Poe blinked, green eyes winking out briefly, then lighting up again. His purr was like a piece of velvet being passed across the ear of the world. The long mirror shimmered in the glare of the tall black candles. Sister Sorrow laughed softly, nestling one ivory cheek against the pulsating animal body positioned on her shoulder. She nuzzled him.

'Poe, Poe,' she whispered, 'never fear. I know how to deal with enemies. Tomorrow, when the moon is full, and madness rides the night sky, he shall be ours! For everlasting damnation and all eternity.'

The lithe cat meowed, then. All at once. Without warning.

A commonplace noise. Banal, prosaic, all too natural.

Sister Sorrow tossed back her magnificent head and laughed out loud. The derisive laughter rolled around the dark room, as if it might shatter the mirror with its volume and wholehearted execution.

' . . . and I shall not need the *eye of newt, wing of bat, and tail of lizard* to triumph over this poor mortal! He is helpless against my power. He will be a sacrifice to the Lord God Satanas . . . '

And so it was ordained.

And so it would be. Satan without end – Amen.

Sister Sorrow would make it come to pass.

As she did all things on earth and in Hell.

Poe, her familiar, who knew every atom of her body as well as every whim and vagary of her devilish moods, arched his feline form, licked his whiskers, and purred almost proudly.

He knew the sound of his mistress's voice, too.

He could sense when she was poised on the threshold of some great act, some masterful sorcery, some tremendous evil.

After all, he was Sister Sorrow's familiar.

Her cat, her companion spirit – her other body.

He had not been named for Edgar Allan Poe without good reason.

Like that tormented genius of the world of books, Poe the cat was all too conversant with the forces of Evil –
– and the ways of the Prince of Darkness.

But above all, he knew Sister Sorrow better than anybody!

'Come, Poe.' She placed her red mouth to his furred ear. 'We must to bed . . . the dance is ended . . . the *rune* is done. . . .'

Poe purred his approval. His pink tongue flicked at her skin.

Sister Sorrow, still clasping him firmly, turned to the tall tapers. Pursing her lips, she blew. Once, twice. In a flash, the chamber was plunged into darkness. Only the incense-laden stench of expired flame and melting wax remained. The floor-to-ceiling mirror vanished. The darkness was impenetrable. Nothing could be seen. A black void had claimed all things. Including Sister Sorrow and Poe.

The Dark conquered all.

Tomorrow was another day, indeed.

POOR BOND IS DEAD

'Ah, Mr Bond. There you are.'

'I hope I'm on time, Miss Copely.' He glanced around admiringly at the lush office of Chic, Incorporated.

'Yes, you are. Promptness is an admirable trait in any walk of life. I see you've brought your sketches too.'

'I don't mean to be pushy. But I still think it would be well worth your while to look at them.'

'Later. But not now.'

'Anything you say. Now, where shall I take you to lunch? I'm an out-of-towner but I do know of some good restaurants. Unless you have some favorite place. Ladies first, you know.'

'Really? Then in that case, I insist. In fact, Mr Bond, I've had a tiring morning. I do seem to have caught up on most of my business for today. There isn't any need for me to come back at all. So, you are fortunate, Mr Bond. And if you will permit me – '

'I'm footloose and fancy free, Miss Copely. My time is my own. You name it. I am yours.'

'How gallant you are. Very well then, you are in my hands. Let's go to my place. We career women do like to turn our kitchens now and then – so that we don't altogether lose the art of cooking.'

'That sounds great. By the way – '

'Yes, Mr Bond?'

'Would you like to see Amy's photo now? So I can prove I just wasn't making a cheap pass yesterday.'

'I trust you, Mr Bond. I prefer to wait until we are away from all this office turmoil and are just two people sitting around a table, making small talk. Is that all right with you?'

'I told you, you're calling the shots. I am in your hands.'

'Well said, Mr Bond. Now, let me get my coat. I've already called à cab, you see. I rather thought you'd see things my way.'

'Why shouldn't I?'

'Why not indeed?'

Warning bells should have sounded and kept on ringing in Philip St George's head, but they did not. Miss Copely, vibrantly stunning in another of her flowing robelike dresses with billowing sleeves, was completely bewitching at twelve o'clock of the new day. The brisk bite of winter air and oncoming snows he had dismissed. It was proper weather for the dead of December. But Miss Copely's dialogue and deportment, as he stood before her secure in his Peter Bond characterization of bespectacled, rather oafish geniality, in the proper ill-fitting tweed suit, was far too formal and a shade too concise for reality. For humanity. The woman was thoroughly theatrical, even allowing for her incomparable beauty and vivid charisma. Stilted, all in all.

Yet perhaps because he was playing a part himself, Philip St George was blind to the falseness of her manner. And after all, he was getting exactly what he bargained for. What he wanted above all things at the moment. The location of the coven overlooking Central Park, which he was equally certain had to be Miss Catharine Copely's own home. Therefore he counted the sudden offer to a tête-à-tête in the lady's domicile as pure serendipity. The machinations of a merciful God who for once might assist the angels instead of the Devil. Never in his crowded career had he ever been more wrong.

He hadn't known of Brother Nicholas, the services he had rendered.

Duane Farmer had done his job superbly. He had indeed been a shadow.

He hadn't reckoned with Catharine Copely's instinctive intuition.

Or magical attributes of divination. Her sheer mind-reading.

So he walked boldly into the spider's parlor. Sure of himself.

Even as the cab she had summoned whisked along Fifth Avenue, mixing with the usual melee of Manhattan traffic, and he was seated in close proximity to the remarkable woman, still the alarm bells did not ring. He was alert, of course, ready to dodge or account for any conversational trap she might spring. Catharine Copely disarmed him with ease. She didn't ask about Amy Bond, nor mention Karl Frobisher, nor make any references to the Bond background, the art designer field. Or anything. She merely folded herself up in a full-length dark fur coat which made her oval ivory face a fantastic mask of perfection. The fur, he observed was a pure Russian sable, and as he studied her he seemed to note for the first time that her fingernails were unpainted, and as carmine as her mouth was, it didn't indicate lipstick of any sort. She did not look at him directly, merely staring straight ahead during what proved to be a rather short ride despite the heavy noon traffic. So he could not read her eyes at all. Much less see them.

When the first few light flakes of snow cascaded down, pasting themselves to the cab's windows, she murmured something under her breath, and he saw that as an opportunity to say something.

'Snow. What do you know?' He laughed the way Peter Bond would. 'Guess we'll get that white Christmas after all.'

Catharine Copely stirred within the confines of her fur coat.

'No. It will not last. And then it will turn black. The city will leave its polluted stain on all of it . . . but black is better – much, much better . . . the absence of color . . . the total starkness . . . '

Her low, vibrant voice trailed off to a whisper.

A simple soul like Peter Bond wouldn't have said anything to something so incongruous as that, so Philip St George didn't. Miss Copely lapsed into silence once again.

Nothing more was said until the cab slowed to a halt before one of the tall apartment buildings on Central Park South. Just down from the New York Athletic Club and Broadway. Philip St George's brows knitted in a concentrated frown. The frown was to mask his satisfaction.

' . . . *overlooking Central Park*,' Norma Carlson had said.

This narrowed that broad generalization considerably. By miles!

He was drawing on target. Taken to the coven's hideaway

by the leader herself. A penthouse with a panoramic, sweeping view of the park, the green expanse lying at Catharine Copely's feet like some mammoth doormat. A lawn stretching all the way up to One Hundred and Twenty-fifth Street, from East Fifty-ninth Street Columbus Circle and Central Park West. This had to be the place, unless Norma Carlson had been seeing things or been spooked, hexed, or drugged.

Unless – a lot of things. He gripped the folio of sketches, firmly.

'Here we are, Mr Bond,' Catharine Copely laughed lightly, extending her slender hand so that he could assist her out of the cab's interior after he had paid the driver. 'You're in my power now, you know. You don't eat my concoctions, then you'll starve.'

'I won't starve.' Peter Bond's laugh matched hers, ever the gallant fool. 'Wait till you see my appetite. I'm part lion.'

Philip St George, escorting her into the building, toward a waiting bank of two elevators at the rear of a tiled, planter-lined lobby, was aware of only one thing as he clutched his folio.

One sensation.

A sharp tremor ran up and down his spine. The nape of his neck tingled. It was the same wave of anxiety and expectancy which had filled him on the last one hundred feet of the climb up Everest. He knew himself very well. He realized that he was up for this cat-and-mouse game he was playing with Catharine Copely, the would-be witch.

Way up. And finely honed to a sharp edge of self-defense.

Perhaps it was just as well. He wouldn't take too many chances with Catharine Copely. For all her theatricality and hokumish demeanor, the woman was remarkable all around. No second-rater, no ham-and-egger, or even a bush-leaguer who didn't even know her own special game. Not a patsy, either. She was too formidable.

Still, despite his mental preparation and his determination not to relax his guard, he had underestimated the lady.

He never should have walked into her apartment without knowing the territory, the lay of the land or the disposition of the enemy forces. But for all his incredible skills, Philip St

George was only a man, after all.

A rare man, it was true, but still all too human.

And vulnerable.

Sometimes the crusade and its demands blinded him.

As it did this time, when he least expected it.

Within a few moments of entering Miss Catharine Copely's home he was to learn, once again, the sad and all-too-dismal truth of that handicap. Satanism slaughters humanity.

Sister Sorrow was more than just a woman who called herself a witch. She was Eve, Cleopatra, Lucretia Borgia, and the Devil's Daughter, all rolled into one. On all six cylinders.

And all on a winter's day with the snow just beginning to fall.

Snow that would turn black, as Catharine Copely had predicted.

As Sister Sorrow had decreed.

The ride upward in the plush, pleasant elevator was memorable, brief as it was. It proved to be the turning point of Philip St George's fall from the self-imposed heights of cold detachment. They were alone in the car, he and the incredibly beautiful sable-clad creature with the deep, lustrous eyes. Eyes with stirring wonder and mystery in their pools.

Catharine Copely leaned back against her side of the elevator car. Her arms were drawn up to her bosom, the tapering, slender hands tugging the fur collar of the coat to her exquisitely ivory face. Philip St George stared into her eyes. The shivers of tension and anticipation had, oddly, all gone now. There was warmth in the close atmosphere. Warmth and an almost dizzying aroma of sweetness. As if so many fresh, crushed flowers were lying in scented holy water. He blinked – and the alarm bell in his brain *did* try to go off. But it couldn't. He was strangely exhilarated, completely relaxed. The tingle in his chest and hands was inexplicable . . . *now*.

'Peter,' Catharine Copely whispered. The name was a prayer.

Her voice, animal, husky, sensually laden with promise and invitation, had made all time stop. The car seemed to float upward. He shook himself, remembering how she had placed a slender forefinger on the black enamel buzzer which indicated P for penthouse floor.

How strange. He couldn't remember ... *when.*

Suddenly that seemed like hours ago. Yesterday, perhaps ...

'Yes, Catharine,' he said. The dark eyes bored now. He wanted to drown himself in their pools. His loins were abruptly on fire.

'Kiss me, Peter ... I know you want to. Don't hold back. Do it ... I want you to, Peter. Now ...'

He kissed her, then. Without will, or thought, or conscious resolution. The large folio of sketches slipped away.

It was like falling into quicksand, descending into a maelstrom. Crashing waves pounded along the shoreline of his mind, breakers boomed, a light exploded. He submerged in her arms. She was all about him now, enveloping him. With her demand, her desire, her hungry, biting need of his mouth. The kisses glued him, melding him to her and he dug his hands insistently into soft, pliant curves, into swells and mounds of glorious womanhood. His mind rioted. And still he kissed her and she kissed him. Their mouths opened, their teeth bit, their tongues locked ... and the car tilted to a standstill. Seeming to careen to a halt. Swaying, rocking, undulating. ' ... *you can have me, Peter ...* '

The incense in the air was like the musk of a woman's bedroom, tinged and heightened with mind-bending, ultrasensory, sexually maddening perfumes. He tore at her clothes, ripping at the long, black, sensuously wound fur coat. The ivory column of her throat ...

Her low, urgent plea echoed in his ear. Impassioned, dreamlike.

Like the crash of a thousand symphonic violins, all clashing.

' ... oh, Peter ... wait a little ... until we are in my own bedroom ... please, darling ... it will be so nice in there. ... '

'Yes, Catharine,' he said dumbly.

That was all he really remembered. ' *... we will make love ...* '

She led him from the car. Through a maze of blacks and reds and purples and oranges – all the radiant hues of the spectrum. There was a dark, high door, a crescent-shaped symbol set in an upper panel of the barrier. And then more darkness, and a spill of amber light suffused with fiery flashes.

He could hear something like Chinese mobiles tinkling eerily, as if a strong wind were sweeping through the star-crossed, comet-filled, fiery universe, sweeping across Catharine Copely's penthouse apartment. *His woman's* universe.

If this incredible place could be Catharine Copely's penthouse apartment. Surely it was Limbo or Purgatory, somewhere between Heaven and Hell. Somewhere in the darkness of the all-enveloping night.

And still she led him forward, holding him by the hand.

Her fingers were like silken webs, spun out fine. Holding, owning, possessing. Transmitting desire and love through the very fire of her flesh. Blindly he stumbled after her. Hungering for her, throbbingly.

Peter Bond or Philip St George, she owned him.

He was hers.

To do with as she saw fit.

All about him insane sights and images and visions kaleidoscoped, skyrocketed, and teetered. The face of Pan here, a hunchbacked cat there, glittering phantasmagoria all over. A tall clock with a luminous silver pendulum, an ancient bell with ponderous clapper, hour glasses side by side containing dark sand, Satan masks on the walls, the floor beneath his feet a mosaic of pentagrams with insidious, carved inscriptions – all a jumble, all a mad, wild assortment of color, size, and malevolent meaning. His head swam but he only had eyes for the woman. Her. Catharine Copely. Sister Sorrow. The woman he loved, the woman he wanted above all others. He felt buoyant and yet as heavy as lead. He felt like he would explode and yet everything had drained out of him. Including will, reason, and a sense of things as they had been.

This was unearthly. This was a new horizon for Philip St George.

This was the other side of the world. Another lost horizon.

This was sheer insanity, not Shangri-La.

'*Peter?*' Catharine Copely whispered.

He reached for her.

He couldn't find her. She had moved, drifting off, like a vapor. Ethereal and ectoplasmic. She sounded far away, too.

'*Here is our bed of love, Peter . . . where you can possess*

92

me . . . and do as you will with me . . . I can't fight you,
Peter . . . I am yours.'

She shimmered before him. Wavering in the half light, the hell's illumination, the shining stars. She was yearning, holding out her long white arms. The coat was no longer between them. The billowing outfit had vanished too. Catharine Copely stood before him. The female incarnate. Sculpted, curved, carved, transfigured for all time in one burst of flowing, stunning nudity. His soul compressed – detonating.

He pawed out for her, groping, clutching, seizing.

And she was no longer there. She vanished all at once.

And suddenly he was lighter than air.

There was no floor beneath his feet.

There was only space.

He plummeted downward, with banshees screaming and bats soaring in hideous, screeching tangents of flight about his falling body.

He dropped like dead weight.

Dropping through the trap door conveniently positioned in the very center of the foyer leading to Sister Sorrow's boudoir. Dropping down into the boxlike, stone-walled cubicle where the Divine Sister could deposit any matter until the time when she decided to deal with it properly. Like say around midnight, when the Mass would be served.

Philip St George was unconscious before he slammed into the stone floor some ten feet below the level of the false-bottom foyer.

As the parquet floor above slid back into place, maneuvered by the lever concealed in the strip of molding which bisected the walls before Catharine Copely's bedroom, the lady herself smiled triumphantly from the threshold. Behind her, a wall-to-wall tapestry of Satan being driven from the Gate of Paradise gleamed in flaming reds and maroons and auburns. Never had she looked more Satanically lovely.

The horned head of the fallen Lucifer, with its malignant and twisted features, however, was a match for her own.

Both expressions were strikingly similar. Mirror reflections.

Fully befitting a Daughter of Darkness.

A devil from Hell, herself.

GATHER, DISCIPLES

Toward sundown, with the midday snowfall slowly yielding to a mixture of rain and snow, many people, many lives, and many incidents, though seemingly unrelated, were moving forward. The pacing of all these factors would decide the fate of Philip St George. It was as if some great unknown race were being run, though few of the participants were aware of the nature of the event. Men and women, who at their best are unreliable creatures, self-interests forever holding top position in their existences, seldom do know what truly will happen next.

Therefore in many a nook and corner of the metropolitan uproar that is Manhattan on any given weekday, fifteen completely different and varied people from all walks and professions could barely conceal their impatience with the torturously slow advent of nightfall.

Some for the very same reason, the identical purpose.

Others for far more human motives.

But all of them had a stage in the life of Philip St George, whether they knew it or not. Their private star was crossed with his.

Wayne Rilling, theatrical agent, was drinking far more Scotch than was good for him, alone in his Madison Avenue office. It had been a hectic day, what with interviewing seven prospective clients for the new Blane Thomas play, scheduled for early spring. Worse than that, the prospect of another session in Sister Sorrow's Inner Circle that very evening had worn Wayne Rilling to a frazzle. He always had to bolster up his nerve to go through a Black Mass. More so than ever now that he had seen (with his own eyes!) Charlie Carmody changed into a green garden snake. Jeezis – how had he ever gotten mixed up in such a crowd in the first place! It was one thing to do it for fun and games, kicks – a new experience – but the heavy stuff was a bit rough, even for a Broadway tough guy like Wayne Rilling. But whatever Sister Sorrow wanted, she got. And that night she wanted Brother Judas to play illuminator. To hold the black candles over the altar – goddamn her witch's soul. Her black witch's soul.

Rilling had another drink. And another, and another.

Bedelia Aaron, dietician, wide-hipped and blonder than a palomino, was looking forward. Her office routine, recommending one low-calorie, fat-free diet after the other to ruthless and rich Park Avenue clientele, had hardly shaken an ounce off her own ample derriere. For one thing, Bedelia Aaron revelled in Sister Sorrow's get-togethers. Nothing, not even sex with both sexes, turned her on as much as watching the sister put the blocks to some chicken-hearted, backsliding member of the flock. Not to mention masturbation with a crucifix. No, Bedelia Aaron was not upset by tonight's coming attractions. Sister Lilith was all set for another main event. One more Black Mass put on by the woman who really knew how to run a ceremony. Sister Sorrow. Truth to tell, Sister Lilith had had an orgasm watching Brother Carmody turn into a wriggling, insignificant green garden snake.

Max Toland, banker and Brother Herod, was more than anxious to strike the gong, opening another of Sister Sorrow's masses. The gray-haired, aging head of the largest bank on Seventh Avenue should have retired years ago and perhaps would have if he hadn't gone to Haiti for a vacation and gotten hooked on Black Magic as a means of restoring his virility and interest in life. Now he could phone his wife in Massapequa, insisting he was too tired to come home and would stay at the club, and not feel one guilty qualm. At sixty-seven, Max Toland was more than ready to join Sister Sorrow in any form of Satan worship she could conceive. He adored Sister Sorrow and was longing for the day when she would be his in the mating circle. So far she had not been his partner. He was certain this was not intentional but merely luck of the draw. He was perfectly willing to wait his turn. Perhaps tonight!

Sloane Gilley, Broadway producer, stayed away from his office all day. The masses took all his strength, so he girded himself for the night's coming activity by using his private sauna, whirlpool, and swimming pool, which were a part of his Tudor Place living quarters. Gilley was completely bald and looked sixty years old though he still had not reached his fortieth birthday. He was one of the most successful producers in Gotham, having the magic touch with the plays of promising young playwrights who also became his lovers in short order. Brother Attila, in fact, who would be deacon on this night of all nights, was not the most secure and happy man he very well could have been had his sexual proclivities been of a different stamp. He scorned the idea of a heaven.

Duane Farmer, Brother Nicholas, slept around the clock, in the loft building where he roamed like a ghostly scarecrow, living a reclusive life among his books, papers, and sacreligious paintings, until Sister Sorrow called. He had earned his rest, thanks to his diligent and rewarding surveillance of the offender, Peter Bond. Being celebrant of the Mass would be all in a day's work to Brother Nicholas. When you are a devout believer, as Duane Farmer was, eking out a remunerative existence as a spot-news commentator, you are very likely to take things in stride. Brother Nicholas would leave everything to Sister Sorrow. What she ordered and directed would be done! There could be no two ways about that.

Sister Circe, who was Alice Ainsworth, already designated as the incense bearer for the Mass, had perhaps the busiest day of all the coven. The magazine she toiled for was a monthly given over to the travel and entertainment field and she was hard put to get all her copy, artwork, and ads pasted to layouts so that the March issue could be air-expressed to the printers by the agreed deadline date. Therefore she went through her day in a frenzy of meetings, last-minute proofreading, and scheduled appointments with the editorial department. But when five o'clock rolled around and she suddenly remembered what was on the menu that evening at Central Park South, a sly smile parted her hawkish, brunette-beautiful face and her heart actually skipped a beat. Sister Circe enjoyed the men in the coven as much as the next girl did. Perhaps more so. If she hadn't been a successful magazine editor, in all likelihood she would have been a call girl or a prostitute. At the very least, she was a confirmed but under-control nymphomaniac.

As for the others, Jason Browne, architect; Buck Benson, engineer; Wanda Ravelli, school principal; Baldwin Simpson, veteran character actor; and Sophie Smythe, commercial artist – all devotedly interested in Sister Sorrow and the coven, for private reasons of religion, need, and, perhaps, the sheer excitement of belonging to something secret, hidden, and *illegal* – all of them stewing all day, following their careers and personal lives until their individual wristwatches and all the clocks and timepieces of the city told them that it was drawing closer and closer to the hour of rendezvous at Sister Sorrow's. The hour for but another *Messe Noir*. The Black Mass.

This night was to be special.

This night, a newcomer. A tyro in the lists.

To be dealt with and served up as a sacrifice to Satanas.

An offender who had dared Sister Sorrow's wrath.

And they all would be ready to take part, do their required task, and serve Sister Sorrow, obediently, blindly, faithfully.

Brother Mars Browne.

Brother de Sade Benson.

Sister Medusa Ravelli.

Brother Hitler Simpson.

Sister Bertha Smythe.

The Inner Circle would be complete.

Sister Sorrow would make up the twelfth member of the coven.

The offender would be the thirteenth.

There would always be time, and opportunity, to locate another brother or sister to make the circle complete again.

There was always time, when Satan was your master.

And God your arch-enemy.

Sister Sorrow had trained her disciples exceedingly well in the art of being fallen angels who no longer followed the dictates of precepts of Heaven. Casting them aside for the concepts and rules of Hell everlasting. The coven trembled in anticipation.

7   97

Whatever their personal reasons, never had any one of them so looked forward to a Black Mass as they did to the one scheduled for that night. Sister Sorrow had promised, with all her evil relish.

The sister always made her promises good.

Made them *bad*, too.

Satan bless her!

The only two phones which could have established some form of contact or liaison with Philip St George that day had proved an absolute dead-end for Sidney Kite's secretary, Miss Irene Walters. Her harried but self-controlled employer had asked her to get in touch with Phil when the research assignment had uncovered a wealth of material and a harvest of goodies which might or might not be exactly what the young millionaire crusader needed. Kite didn't know, but he wanted to find out right away. Miss Walters had worked long and hard, justifying her boss's good opinion of her once again. But the phones at the Explorers' Club and the one in the sky-high roost in the middle of Manhattan elicited no reward.

A calm voice at the Explorers' Club assured Miss Walters that Mr St George had neither come to the club or phoned in any messages. Miss Walters left word for him, as any good secretary would, and then tried his apartment phone for the tenth time that active day. Again, no response. When the answering service belatedly cut in, the information was the same as the Explorers' Club. No Philip St George. Not a word from him.

Nobody, of course, knew about the Peter Bond registered at the Americana. A good cover can have its drawbacks sometimes.

Miss Walters hung up for the last time in exasperation and then walked into Sidney Kite's office to relay the nothing news.

'He's off somewhere, Mr Kite. Or dropped out of sight again. I'm sorry, but that's about the size of it.'

Sidney Kite snorted from the depths of the wide window where he was dejectedly surveying the hordes of ant life streaming by below. That was something that hadn't changed in forty years, either! Why should Phil St George be any different?

Young damn fool! A genuine *kibitzer*, that's what he was. Messing around in things he had no business sticking his good-looking nose into! Devil worshipers – cockamamie trouble-makers. Nogoodniks. Let the police deal with them.

He turned wearily from the window, aware that Miss Walters had lingered in some sort of show of concern. He smiled at her. He liked Miss Walters. Too bad he wasn't a womanizer – Walters was first-class.

'You shouldn't let him worry you so much, Mr Kite. Honestly. He's not exactly underprivileged, you know. With his money and brains and those good looks of his – well, all I can say is I just can't feel as sorry for him as you do. I know about Lake Placid, but that was long ago now. And he's got his whole life ahead of him – I hate to see him wearing you out like this. You're a good man, Mr Kite.'

Few things ever surprised Sidney Kite. Miss Walters' sterling loyalty and outspoken criticism of client number one did. His eyebrows went up and his doughy, large-nosed face showed that surprise.

'Irene – I didn't know you cared.'

She managed a sheepish smile and shook her head.

'Don't tease me. You know how I feel. And you know what I mean. We did our part. We got him what he asked for. Now it's up to him to come pick it up. Or should we deliver it to his apartment by messenger? Heck, he's a big boy now, Mr Kite. He can take care of himself.'

Sidney Kite suddenly sat down behind his desk, feeling very old. Walters was right. Why should he carry the weight of Phil around on his back? Father-and-son stuff was hooey. To each his own. Yawning, he probed at his tired eyes, all too willing to buy some of Miss Walters' common sense. It was one of those days, snow in the air, bitter cold on the outside, when a man could feel every one of his fifty-odd years.

'Ah, maybe you hit it on the head, Irene. Why should I beat my brains out for him? He'll pop up in a day or two, asking for this junk, and then again maybe we *could* send it to him. I don't know. Let me think about it awhile. Trouble is I've been look-ing over some of the material you typed up, along with the newspaper stuff, and – '

'What about it, Mr Kite?'

Sidney Kite shrugged. 'There's some real important stuff you got there. Facts which didn't hit all the papers. Missed it myself. I wonder if Young Galahad is on to that part of it. If he isn't, he could be operating with one arm tied behind him right this minute. Whatever he's up to. You know what I mean?'

'How can I,' Irene Walters protested, 'when I don't know what you're talking about?' She fingered her blonde hairdo self-consciously.

'This,' Kite grunted, poking a stubby forefinger in her general direction. His sudden keen expression was the one that the perjuring witness usually saw in a packed courtroom. 'I know enough about murder when it comes in bunches. Like all these dollies getting beheaded. Any cop worth his salt looks for what they call a *common denominator*, something that would make all these women have something in common even if the record is pretty clear that they never once met, grew up together, or even worked in the same factory once upon a time. A common denominator which would have made them the target for this kind of killing spree. You following any of this, Irene? It's easy when you know how.'

'I – I think I know what you mean.' Miss Walters nodded gamely. 'What you're saying is if all the women were Protestants and the killer hated Protestants or they all had green eyes and he hated green-eyed women because his mother was green-eyed and used to beat him as a child – that sort of common denominator. That it?'

Sidney Kite beamed, as tired as he was. Walters had scored again.

'On the nose. 'Course, here we got all kinds of common denominators. These look like cult killings, one of Phil St George's cockamamie Devil-lover clubs. And the girls are kind of all orphans, unmarried, and loaded with the coin of the realm. But it's more than that. And I found it and I'm sure Phil hasn't had a chance to spot it yet.'

'Go on,' Walters urged, almost commandingly, forgetting her place. 'I don't remember anything particularly startling while I was going all over that material.'

'You didn't know what to look for,' the lawyer said almost

100

sadly. 'I did. And here it is in a nutshell: the Misses Taylor, Richardson, Black, Busby, and now you can even include this last one, Davis, and are you ready for this, Irene? – *They were five-footers weighing around one hundred pounds. No more, no less. In short, they were all tiny.*'

'But that's so – so awfully commonplace, Mr Kite.' Miss Walters couldn't hide her disappointment. It was rather as if she were expecting a far more sensational clue, really hot and juicy.

'Maybe it is,' Sidney Kite rumbled, 'but it's also too damn coincidental and totally against the law of averages. Five out of five? A thousand per cent? No way, Irene. It has to mean something, has to be worth knowing, and personally I wish Phil St George knew it as soon as possible. It could make a difference.'

It would have, too.

If Philip St George had known, perhaps he wouldn't have wandered quite so cavalierly into Catharine Copely's midst.

Perhaps, a lot of things.

The falling snow was still fighting its losing battle with the rain and wind as Sidney Kite waved Miss Walters back to her desk and tiredly returned to some pressing matters relating to the business affairs of Kite, Dorn, and Schindler. Law was still his first love.

There was no communication from Philip St George the rest of that working day, despite Miss Walters' best efforts and Sidney Kite's habitual concern. But disappearing acts were nothing new with Phil.

Sidney Kite somehow sensed it was always going to be that way, too. Damn all do-gooders everywhere! Always sticking necks out.

What was the sense of asking for information if you didn't bother to call up and find out if it had been obtained? If it was ready?

Did Phil expect him to do everything? Quick? Like a bunny?

Well, Mr Crusader could go soak, and good luck to him.

Wherever he was.

Unhappily, the kind of places and tight spots a Philip St George could get into were exactly the sort of things Mr

Sidney Kite didn't care to think about. Such thinking only made a man feel very much older. And out of the mainstream of life. And useless..

Sidney Kite didn't like to feel that way.

Who does, really?

Even cockamamie crusaders were better than that.

Book Three:

HELL'S HIDEAWAY

'When you are happy, truly happy, you want to share it with the world. With your friends.

Life was meant to be shared.

Not Death.

Leave that to the Devil.'

> – Death Is a Dark Man, *by*
> *Dora Highland,* © *1974*

SHE'S A WOMAN, TOO

Catharine Copely perhaps was busier than anyone else that fatal day. From the moment she lured Peter Bond into her incredible parlor, high up in the Manhattan skies, she had worked long and hard, albeit fluidly and with scarcely a wasted motion.

All of her characteristic grace and lithe physical movements served her in good stead as she disposed of the remains of Peter Bond, bogus fashion designer and man of mystery. And who knew what else? Though Catharine Copely flattered herself that this foolish intruder was nothing more than a private investigator of some kind come to find some clue or evidence of Copely guilt as pertained to five cult sacrifices. Though how any of the beheadings could have been traced to her was beyond her understanding. Yet, learning what Mr Bond did know – before he fulfilled his role as altar in tonight's Black Mass – was precisely what Catharine Copely intended to discover.

Before the others arrived. The coven need not know about Mr Bond or any of his previous history or background. Or sins!

Suffice that he would be sacrificed. And serve the Prince of Darkness as living offering to the everlasting God of the Pit.

There is a great deal to do when you lure a man to your apartment and then drop him into a trap-door, false-floor arrangement especially constructed for such a purpose. Webs demand industry.

Catharine Copely did it. With the familiarity of long practice and countless previous experiences. And through relish.

First she retrieved the large folio of sketches from the plush

105

elevator, then pressed a bracketed wall switch which sent the car back to its starting point – the lobby floor. She swept into her domicile, moving like a phantom. As she glided she removed her long fur coat and dropped it to the floor. The swirling, billowing robe of the costume she wore rustled with silken rhythm. Catharine Copely's vividly dark eyes glowed as she set about dealing with Peter Bond.

She did not flick on any artificial illumination, preferring to work by the pale grays and whites of the remainder of dying daylight which cast ghostly shadows across the parquet floor. Her tall, shapely silhouette might have been a true spirit. A shade.

In the heart of the peculiarly fashioned foyer just before her boudoir door, she once more paused beneath the hanging tapestry of Satan in exile and fingered another hidden wall switch. Then she stepped back, away from the floor space, within the arched confines of a doorway. She stared down at the hardwood floor, a strange light in her eyes.

Almost instantly there was a whirr and steady rumble of machinery of some kind. The floor split in half; both wooden sides parted like connecting elevator doors moving back, and were flush with the walls of either side. Darkness and emptiness lay at her feet. But not for long.

Magically, as if levitated by Dark command, the crumpled form of Peter Bond was raised into view. Continuing to travel upward, sprawled inert on the stone floor of the boxlike cell into which he had been dropped. But now the floor had become a stone table, an altarlike repository which did not stop moving until it had cleared the level of the foyer and reached Catharine Copely's middle. The machinery stopped rumbling. Gears and cogs went quiet. And now Peter Bond lay on his stone bed, unconscious, centered as if on a funeral pyre in the dead center of the foyer. There was perhaps four feet of clearance to either side of the stone table. At its head lay Sister Sorrow's boudoir door, which was closed. At its foot stood the hall corridor. To the eastern side was Sister Sorrow, poised like a queen, studying the prone man before her. On the western side was the blank wall with its strip of maroon molding bisecting the corridor height. The tableau was eerie.

Sister Sorrow removed her billowing garment, allowing it to trail to her ankles. She stepped from the rustling heap and raised her long, ivory arms outward, as if flexing them. Her magnificent figure, now covered by a wisp of material across the bosom and hips, flaming red scarfs which might have been chiffon or silk, made her an incredible apparition indeed. In the dim light of the foyer, with her long black hair falling, she was like something out of a fantasy. An artist's vision.

Peter Bond was in no condition to appreciate the astounding female poised above him. His eyes were closed and his breathing was fainter than a dying whisper. Sister Sorrow's eyes glinted as she bent over him. Her slender, expert fingers made quick work of the clever characterization and makeup job with which Philip St George had made himself into Peter Bond. But now Sister Sorrow knew what to look for.

She found it, too. All of it. With a mounting admiration, an almost grudging tribute to the fool who had sought to trap her. Not for a moment had she thought that Peter Bond was *not* wearing his true face and physical demeanor. It was only through the report of Duane Farmer that she had suspected anything false about him. The strange behavior.

But this – this deception – the surface one at least, was truly uncanny. Worthy of the greatest thespians in the world.

One by one, she made a neat pile of the integral segments of the Peter Bond identity. They added up to a considerable total.

There were the spectacles, of course. Superficial, but good.

And the carefully placed wads of dental wax in each jaw to change the lean pattern of the face. Then the built-up putty bridge on the nose to alter that, too. The flesh dyes, when rubbed completely from the face, produced an amazingly bronzed and unmarked complexion beneath. And the hair – the muddy, brown, limp, dull colors and textures did not quite prepare one for the full, bushy shock of jet back hair with the most unusual silver streaks at the temples. Slightly impressed now, Sister Sorrow worked on, moving even more rapidly, eager to learn about this intruder.

She undressed him without delay. Without modesty.

The ill-fitting, coarse tweed suit fell away, the brogans were discarded, the underclothing tugged free. Sister Sorrow's amaz-

107

ing eyes widened in greater surprise. Dye still clung to the hands but now she could see for herself the amazing physical specimen revealed on the stone table. This man was no mere mortal. He was tanned and powerful and beautifully fashioned as any statue in a museum. A copper Apollo, a mute and magnetic male giant out of a Michelangelo fresco. A divine Adam, a true warrior – a Titan. Sister Sorrow's gaze roved down to his loins.

There was a sudden catch to her throat, as if she found it difficult to breathe. The narrow foyer echoed with a tremulous, almost feminine sigh. Sister Sorrow was abruptly astonished by her own reactions to the mere sight of a nude man. Something she was as familiar with as her own lovely reflection in a mirror.

But this man – this intruder – whoever he truly was – was gifted in every conceivable department. He was superbly endowed. Pan would have envied him for his incredible appendage. The principal male tendon was a thing of beauty, even dormant and idle. *The Ram's staff!*

Sister Sorrow shook herself and turned her attention to the contents of Peter Bond's pockets. The billfold, loose change, and a set of two keys she ignored. Just as she was not convinced by the wallet of plastics IDs; the licenses, credit cards, and Social Security number were all too easily forged and obtained. Peter Bond would have covered his tracks well on that score. Those things were certain to be standardly checkable. But the manila envelope she found tucked in the inside breast pocket of the tweed jacket drew her cautious survey and study. Again, that low velvety sigh escaped her. A sound of tribute.

The photo of Amy Bond, impossible to detect as a fraud, was a convincing cheat. Yet Sister Sorrow instinctively knew that this man had somehow obtained a photograph of her and doctored it accordingly. If his genius for disguise and characterization could have so well masked a Greek god prototype so that he looked to the world like a genial, bumbling, undistinguished oaf, a trick photo should have been child's play. At the very least. Yet – who was this man?

Sister Sorrow did not know. She who knew nearly all things.

She stared down at him for a very long time, struggling for

the answer. Surely a specimen that looked like this had to be *somebody!*

Somebody select and very, very special. Olympian in achievement.

Never, not in all her career as witch and observer of foolish mankind, had she ever set eye on such a striking male animal.

The face was remarkably handsome, chiseled out of flesh and bone and sinew to seem like the representations of deities on ancient coins. The physique was purely classic, broad-shouldered, narrow-hipped, long, and superbly proportioned, the sort of figure which would seem as proper in loincloth or armor as any of the male costumes of history and the universe. Naked like this, it was breathtaking, altogether intoxicating. Sister Sorrow was a witch but she was also a woman. A splendid pagan goddess who never had had a god of her own. Duane Farmer, Brother Nicholas, paled like a thin burlesque scarecrow of a celebrant when compared with this Mr Peter Bond!

Sister Sorrow's eyes narrowed. Her exquisite face tightened into an expression impossible to fathom. No one who knew her, no one, could have guessed or divined what she was thinking at that exact moment, as she stood alone with the naked giant on the stone table before her in the trap-foyer of her own apartment. Breathing in a curious way.

She seemed to have no qualms about the tremendous male suddenly awakening and trying to fight his way out of her clutches.

There was no wonder in that. No risk of any kind at all.

Peter Bond had never had a chance. Not even Peter Bond who was really Philip St George who had underestimated the witch of the coven overlooking Central Park South. He had calculated badly.

Her hypnotic powers were extraordinary, it was true. And she had truly mesmerised her victim in the quiet elevator as it rose from the tiled lobby. But long before that, Catharine Copely had prepared her conquest with all the witchcraft and modern magic that any clever lady Satan would. Philip St George had been completely oblivious of the method employed to take him in. Though on his guard, he had never expected trouble from the quarter from which it came.

He had been expecting danger and plots and tricks at the apartment house; he had not thought about the cab or the short ride to the lady's lair. Unfortunately, that was exactly where disaster had first overtaken him. Sitting next to the exquisitely enchanting, fur-coated creature in the rear of the taxi. In effect, he had been cheek-by-jowl with his own downfall. Yet, Waterloo had never been more lovely. Or fascinatingly feminine and mystically oblique. And poisonous.

Catharine Copely had simply shifted her position slightly as the vehicle navigated a turn on to West Fifty-ninth Street. Her body had pressed briefly against Philip St George's thigh. Lost in the quick, almost casual meeting of their legs was the tiny pinprick of reflex sensation caused by the contact of the barely visible needle cleverly placed under one of the long fingernails of Catharine Copely.

Not lost was the effect of the drug thus administered.

A mind-bending, disorienting, powerful concoction whose usage by agents engaged in espionage and other subversive projects was known to Philip St George in his crusade career. But the knowledge would not have done him any good anyway. For there was no prescribed antidote for a drug whose combination of sodium pentathol and a new deadly chemical called relalidone generally produced one full hour of unconsciousness; a time-lapse which could be broken into should the administrators care to question the individual drugged.

Or, not at all.

It had been fairly simple, following the injection, for Catharine Copely to turn on her magnetic charm and hypnotic influence full force to heighten the effects of the drug. To bait the trap.

Phil's topple to the stone floor below had completed the knockout sequence. The Copely habitat, with its weird trappings, fantasy adornments, and utterly alien sounds, had merely implemented the whole bizarre scene. The universe had gone off, half-cocked yet full-tilt, into nightmare. Even for a Satan Sleuth.

Catharine Copely now placed her cool, slender hand on Philip St George's chest. Suddenly, she drew back her fingers in alarm. A grimace contorted her beautiful face. Her own flesh

110

burned, as if she had reached down and touched a blazing poker iron, searing herself.

It was – astounding. Like the scourging fires of Hades.

Sister Sorrow sighed again. She trembled, feeling reborn!

The man's body – this godlike creature – exuded the flame of life and throbbing immortality. His body was warm, warmer than any man's she had ever touched. It was as if a river of hot blood and torrid passion coursed eternally through his limbs, to all the extensions of his magnificent form. Sister Sorrow shuddered slightly. A tremor of something, an altogether long-forgotten emotion stirred within her. The palms of her own hands were suddenly moist. Her bosom rose and fell abruptly, as if she had sucked in a lungful of oxygen, all unwittingly. She stared down long and hard at the bronzed giant laid out before her. She shook her head again, long hair dancing down her statuesque shoulders. Her ivory skin glowed a fine sheen.

Bewilderment showed on her face. Almost utter confusion.

And something secret, and unhallowed, like a wicked smile.

Satanism flashed in her eyes.

In that moment she was Lucifer's Daughter incarnate.

Vividly evil, ultimately Devilish, paganly wanton. Gleaming.

Tossing back her head, she raised her face to the dark ceiling and laughed. A low, throbbing, femininely perverse laugh.

'O Satanas!' she murmured, her head still poised, searching the darkness above her. 'Is this He? Have you sent him to me at last . . . the true mate . . . a fitting Pan for your devoted daughter? O Father . . . tell me . . . Prince of Darkness . . . so that I may leave my endless Sorrow . . . and become Sister Joy . . . as you have promised me. Satanas! . . . Speak . . . Give me the word and I will heed. . . .'

There was no reply from the dark ceiling.

Only a rolling, gobbling, chuckling kind of echo. Of high glee.

A sound like an army of mice scrabbling in the walls, perhaps. The shadows in the foyer lengthened. Sister Sorrw lowered her head, her face fell, and the long black hair lay like dark water-falls cascading past her shoulders. The man on the stone table did not move, either. Not so much as the smallest sound escaped him. His breathing, what there was of it, was inaudible. Sister Sorrow's arms went up blindly, her hands splayed out, falling

111

across the massive chest of Peter Bond. Again, contact proved electrifying and pulsatingly sudden. Sister Sorrow shuddered inwardly, her shoulders convulsed, her breasts bobbed. But she kept her fingers fastened to the man before her. Still without raising her face from the floor, her hands began to move. Searchingly.

Roving, traveling, making small counterclockwise circles, then gradually larger ones, until she had touched every inch and portion of Peter Bond's anatomy. She paused only briefly at the loins before her hands continued on. Finally she raised her head, opened her startling eyes, and pushed her face closer to the man on the stone table.

She lowered her mouth, the carmine lips fanned out into a biting, teeth-revealing grimace. Peter Bond's immobile face was but inches from her own. Sister Sorrow leaned over him, her arms resting now on the deltoid muscle of each of his shoulders as she drew her mouth to the corded muscles of his bronzed throat.

Sister Sorrow's eyes glowed like hot coals sparking.

Her pink tongue darted, flicked. With cobra's speed.

Just as the foyer suddenly winked and exploded with light. A purplish, crimson, lambent volley of illumination going on and off in a rapid, nonstopping signal of some kind. Like strobes gone mad.

Sister Sorrow cursed deeply in her throat and straightened up. The throbbing of her breasts was the only sign of her inner agitation of soul and purpose. Witch or not, she was a woman, after all.

The flashing purple and red lights had only one meaning in Sister Sorrow's sanctum sanctorum. One dread purpose.

The coven had arrived.

Another performance of *Le Messe Noir*. The Black Mass.

The very special one which she, Sister Sorrow, had promised them only the night before. In all her malevolent majesty.

Sworn she would give and execute for them, with this very man, lying before her unconscious, as the offering.

The sacrifice. The tribute to Lucifer, Prince of Darkness.

This Peter Bond, this godlike creature who would be punished and scourged and beheaded for his disbelief. For being the enemy of the Lord God Lucifer.

Catharine Copely, the designate, Sister Sorrow, moved away from the stone table and glared up at the weird lighting revolving all around her. Her tall, splendid form trembled with a fierce mixture of seething emotions. The turmoil within her made her sway where she stood. But the firm resolve of her bewitching features and the unholy revelry shining in her deep-set luminous eyes would have terrified any member of the coven had they seen her at that moment.

That total, complete, consummate instant of High Evil.

Yes, the coven would have its Black Mass.

By Satanas, it would!

But a ritual such as the coven never could have deemed possible.

Or anything approaching the norm of such a ceremony.

Sister Sorrow laughed again. A braying, triumphant peal of secret mirth. If the man on the stone table heard, he gave no sign.

The laughter rolled back from the walls of the strange foyer in waves of answering glee.

As if the Devil himself were laughing too.

Joining his daughter in a private joke.

A very private one indeed.

BROTHERS AND SISTERS
IN SHADOW

They came, once more, as they always had.

Like a phantom army, a hidden society, an occult conclave committed to something that demanded the utmost of their cunning and dissimulation. After all, they had to fool the outside world, that other world which would stamp them out and destroy them if it only knew of their existence. So they arrived at the tall building on Central Park South, walking, coming by commercial and private vehicles, never together, singly, and at the proper five-minute intervals. It would take a full hour for the entire coven to be gathered in total number but it would be worth it. No one would suspect, no one would take notice of an assorted group of individuals, of all shapes, sizes, weights, and gender, entering a fashionable building facing Central Park on a gloomy, miserable night in which a conglomeration of snow and rain had united to make for one wretched day all around. A real downer.

December had proved to be the very worst month of the year.

So they came not in a rush, but in a steady, seemingly never-ending stream of humanity. The doorman of the building had seen them all before and never remarked upon their collective identities because Manhattan was a place for party-tossing, group-gathering, and general social groupings from every walk of life. It meant nothing to him beyond the fact that he would nod and smile greetings with more frequency than was usual. The mysteries of Catharine Copely's penthouse retreat, if that indeed was their ultimate destination, meant nothing else, either.

The doorman was a cipher who did little more than play guardian at the outer gates of the building – the street door at the front. The doorman's salary wasn't exactly conducive to services above and beyond the call of that onerous and dull duty. Union or not.

So they came – as Sister Sorrow knew they would.

One by one.

In varying degrees of excitement, tension, anticipation, and fear.

Each appearance was rather like a signature of the personality who entered the tiled, planter-lined lobby and approached the bank of two waiting elevators. The lobby was more often deserted than not.

The theatrical agent, Wayne Rilling, walking unsteadily thanks to overindulgence of Scotch, was short, portly, and bundled up in dark toggle coat with Russian beaver hat to match. Brother Judas's ruddy face was lit up enough without him playing the role of illuminator.

Bedelia Aaron, the dietician, cloaked from head to foot in a sloppy white trenchcoat, sloshed across the tiles in calf-high matching boots. Her blonde hair was tucked under a crimson *babushka*. Sister Lilith's pulse and abundant curves were already trembling in anticipation of the bloodbath that Sister Sorrow had promised.

A mortal sacrifice to His Satanic Majesty.

Old Max Toland never looked more the part of the retired banker, though of course he was as unretired as ever. His tall, spare, angular frame, decked out in a handsome beige topper with pork-pie hat, dog-headed umbrella cane, and ankle-high calf boots, was entirely in keeping with his respectable and lofty station in life. Brother Herod's thoughts, however solemn and aloof though his craggy, leathery face was, were running riot with the possibilities of at last getting his hands on the divine contours of Sister Sorrow in the mating circle, after the sacrifice of the altar to their Lord and Master, Lucifer.

The broadway producer, Sloane Gilley, corpulent, bloodless, and hairless, bald head gleaming hatlessly from the falling rain and snow which had clung to it in the few moments it took to alight from a cab, walked mincingly toward the elevator. His

115

suit was two-tone, rather flamboyant, and altogether wrong for him. That didn't matter, however. Attire, proper or otherwise, was not Sloane Gilley's raison d'être. Brother Attila sincerely hoped that Sister Sorrow's leading character in the evening's little drama would prove to be someone very attractive and thrilling. Too many of the recent sacrifices had proved to be very dull, very uninspiring people. No class or color to any of them, more was the pity.

Duane Farmer loped into view, as funereal as ever. Gaunt, thin as a scarecrow, looking like the complete mortician and Judgment Day all rolled into one six-foot-three-inch package. He was wearing the precise costume he had worn the day before. Henry Higgins hat, blue blazer, gray slacks, boots, unpressed trenchcoat. Sartorial elegance was the least of Brother Nicholas's concerns. A celebrant, as second-in-command, has far more important matters to think about.

Sister Circe, unable to control her fever of impatience or the incipient sensuality in the pit of her stomach, flounced into the stylish lobby on clicking, ungainly high heels, in swirling peasant skirt, cashmere sweater, and black beret, all surmounted and covered with a shoulder-to-ankle cloak of deep blue color. Alice Ainsworth's lean, vulpine face was overrouged, overlipsticked, and her eyes were bright with marijuana intake. She had been smoking grass for about an hour in her Thirty-fourth Street apartment before summoning a cab to take her to Sister Sorrow's. Sister Circe had always found marijuana a sure way to curb her jumps before the Black Mass got underway. Rather like a tranquilizer, grounding butterflies.

So – Wayne Rilling, Bedelia Aaron, Max Toland, Sloane Gilley, Duane Farmer, and Alice Ainsworth. Six covenites all in a row, to be followed by the others. The remaining five, given the count of tonight's order of the ritual. The circle must be rounded out.

And the others came, too.

At the desired five-minute intervals.

Bringing with them, perhaps, the identical feelings, emotions, and/or motives as the ones before them.

Whatever the basic differences, the coven was all the same in its attendance and devotion to Sister Sorrow's Inner Circle.

116

Architect Jason Browne showed up. Medium-sized and middle-aged.

Engineer Buck Benson did, too. Freckled, overweight, blue-eyed.

And then school principal Wanda Ravelli. Petite, dark, angelic.

Veteran character actor Baldwin Simpson was never late. He was right on time, this night of all nights, too. Lean and hard, old.

And last but by no means least, commercial artist Sophie Smythe hove into view, her big, broad body stalking toward the elevator which had been so busy going up and down. Smythe was bovine.

The coven was complete. All present and accounted for.

Ready on the penthouse level of the tall, fashionable building.

Eleven, plus the Sister and the offering. To make thirteen.

And again, as they always did, they entered the long, dark hallway, the corridor so curiously illuminated with flickering weird light, hung up their clothes, one by one, and secured the hooded black robes which each and every covenite must wear until all ceremonies and rituals were completed. Then would come the unveiling, the exposure of carnal bodies in even more carnal pusuits of the flesh. Until that time, no one spoke, no comment was made, nothing in the way of small talk even attempted. Sister Sorrow was the only one permitted to speak in this house. Until she gave the word, no one must utter so much as a sigh. It was the way, no, the rule, of the coven. To speak was to invite wrath, disaster, and possible expulsion from the coven.

One by one the hooded assemblage filed into the large room beyond the long, dark hall. Shadows touched, mingled, bodies brushed together, but not a sound was heard. No human voice gave utterance. The tools, the equipment, the accoutrements for *Le Messe Noir* would be dispensed by Sister Sorrow. Here she was All. She was empress and mistress and holier than holy. It went with the title, the crown, the honor of Daughter of Satanas. Her word was law; holy writ.

Brothers and Sisters blended together in one wide, wavering mass of shadow. Herod, Judas, Attila, Nicholas, Mars, de Sade,

117

and Hitler joined hands with Lilith, Circe, Medusa, and Bertha.

There was a round black table in the center of a room whose walls were deep crimson, whose windowless reaches held nothing but shadows and distortions. There were no chairs. Each of the hooded acolytes moved to a prescribed position around the table, standing where they had always stood, heedless of vacancies and unfilled places.

A door opened somewhere; a wide rectangular spill of amber light bathed the hooded circle in a sudden glare. A blaze of revelation.

Then the door closed but the light did not extinguish.

Another robed figure now stood at the head of the dark round table. The coven lowered its collective head. Two black candles, held out at cloaked-arm's length, revealed the cowled glories of the most divinely beautiful woman anyone in that room had ever set eyes upon. Sister Sorrow's magnetic beauty radiated in the candlelight, caught in an oval of incredible, sensual ambience. The sister spoke then, for all to hear. As she always did.

And said exactly what everyone had been waiting for.

'*Do we dream*,' Sister Sorrow intoned, all of her rhythmic, compelling voice filling the room, reaching all in her service, '*or do we now vote for the death of the offender?*'

If there were a judge, a jury, and an executioner, it was obviously Sister Sorrow who assumed all three roles. She was justice here.

Sister Sorrow called all the shots.

Her home was not only her castle, it was the victim's cemetery plot. The place of execution for malefactors and disbelievers.

And now was the time to speak. The silence was broken. By the only entity who could break the silence of the house. The Daughter of Lucifer. Sister Sorrow, high priestess of this coven.

'Death to whosoever dares to mock us,' Brother Nicholas said.

'Death,' agreed Sister Lilith.

'Death,' snarled Brother Herod.

'Death, death,' Brother Judas murmured drunkenly.

118

'Death in the Black Mass,' rasped Sister Circe. 'Tonight!'

'Death to the offender,' Brother de Sade laughed. 'Now!'

And then all the others chimed in, chorusing, their low, fervent, eloquent voices in effect signing the death warrant of someone they did not know, had never met, and perhaps had no truly valid reason to condemn. Or hate. Or want to harm.

'So be it,' Sister Sorrow said with rich venom and pride. 'Death to he who would defy our Savior and Master. As it has been spoken, so let it be done.'

Her fantastic figure, caught in the flaming circle of the twin black candles, cast an unwavering silhouette on the wall behind her.

She now resembled, if anything, a large monster bat. Hideous, distorted, blotting out the clarity of reason. Obliterating light.

Even Poe, her familiar, had he been present, might have sped squalling from her midst, black spine arched in terror.

Sister Sorrow was awesome when she was like this.

The most evil apparition of them all.

Yet chillingly, incredibly beautiful.

Someone else in that hooded conclave, would have run screaming from the dark room, too. Horrified beyond all reason and belief.

Duane Farmer, dear Brother Nicholas, would have lost the beatific expression on his gaunt, somber face if he could have but read Sister Sorrow's devious feminine mind. And fled in all possible haste.

If he could have known what the Divine Sister was planning for him, as she towered before them all in supernal glory. Smiling. Laughing, extending the flaming candles like beacons in a wilderness.

Replacing him as celebrant of the Mass was uppermost in Sister Sorrow's thoughts. Forever and for all time. The sooner the better.

She was devilishly certain that she had found a replacement for Brother Nicholas, whose ugliness was no longer to be borne. Not even in the black name of Lucifer and the Pit. And all ungodly things.

Sister Sorrow had discovered her living god, at last.

He had been sent by that Evil Majesty who oversees the

affairs of all his subjects. And takes care of their most basic wants.

The sleeping male beauty on the stone bed would be her new mate. Her emperor in life as the Devil was in death.

Philip St George – who had come to her as Peter Bond.

And now must never leave her side again.

Not ever!

It had to be the will of Satanas.

Nothing else would have made any sense to Sister Sorrow.

Two vital factors, unrelated, one natural and the other completely accidental, played key roles in the eventual outcome of the events of that astounding evening in Catharine Copely's hideaway from the civilized world; a milieu which any of her employees down at Chic, Incorporated would not have dreamed possible in a million light-years.

Philip St George was the grateful beneficiary of both those crucial factors. One he had earned, the other was a gift from whatever gods look after people involved in selfless crusades.

The first factor, both natural and earned, was his superb physical condition. For years he had enjoyed all the blessings of fine health and remarkable powers of bodily strength and coordination, a circumstance he maintained through daily exercise designed to keep the status quo. He had also been born with a very high threshold of pain and what was excruciating agony for the average man was no more than nagging discomfort to him. So the upshot of all that was his rather quick awakening from the drug administered to him by Catharine Copely in the taxicab. In fact, he murmured awake only ten minutes after Sister Sorrow had left him. Awoke in the gloomily dim foyer, still resting on the cold stone table centered in the bizarre confines before the Sister's closed boudoir door. For reasons of her own, whatever they might be, she had not depressed the hidden wall lever to return the stone altar-like bed to the boxlike tomb below. Philip St George reorientated himself very slowly, blinking to full wakefulness, finding the terrible tapestry of Satan in exile blazing down from the wall before him.

There was a dull throb to all of his senses but beyond that

120

was only a general feeling of lassitude and well-being, despite the predicament which a full rush of memory now brought back to him. In waves of self-reproach, wonder at the woman known as Catharine Copely, and bewilderment as to exactly how she had accomplished his downfall.

He remembered the cab ride, the elevator, the euphoric spell, the curious eroticism of the beautiful lady, the scent of crushed flowers, the hypnotic, compelling eyes – and then falling, falling. Falling like a ton of bricks for the oldest gag in the universe. A lovely woman turning on the steam heat. In spades. With special effects.

He shook off the bitterness, closed his mind to all that, and concentrated on his surroundings, rapidly totaling up the odds. The witch lady hadn't been all that impractical, relying on drugs of some kind and dizzy spells. She wasn't that foolish, it seemed.

Thick velvet ropes lashed his ankles together, his hands were similarly treated, trapped behind him. Also, he was naked, and it didn't take seeing the scattered pile of Peter Bond's clothing and possessions on the floor in the corner to verify that. Gingerly he washed his tongue around the inside of his mouth. Jaw plates gone, spectacles removed; he wrinkled his nose and knew also that the putty bridge had been wiped away. He restrained a low snort of disgust with himself. He knew what he had to know. Catharine Copely had unmasked Peter Bond. Peter Bond wasn't worth a cent in camouflage, now. The cover had been blown sky-high. It was Philip St George who was Sister Sorrow's prisoner. Not a make-believe man suitable for any occasion of deception. Ruefully, Philip St George tried to think. Coherently, lucidly, and quickly. He had to produce something of value. Like getting out of the velvet handcuffs and ankle bracelets before someone came back to see how he was doing. What he was up to.

He was sure he wasn't going to be left all by himself too much longer. If he had any doubts about that, they were dispelled by a sudden, low, yet altogether solid chanting of voices. A Greek chorus of sorts, raised in some form of prayer or incantation. Close by. As if but a room away. The dark foyer throbbed with the rhythm of those voices. Male and female, intoning in unison. He writhed against his bonds, eyes flying to the scattered remains of the Peter Bond characterization. The clothes. He had to get to the clothes. The walls echoed.

Sister Sorrow might know a lot of things but she couldn't know what a storehouse of weaponry and defenses those seemingly innocuous garments were. For one thing, all the plastic buttons on the tweed coat were cleverly designed capsules containing enough explosive material to blow up the building. The Bond belt, under its leather surfaces, concealed a flexible spring steel blade of the finest Toledo make. Good enough to cut through metal. Bond's shoes, the awful brogans, had false-bottom heels which were the repositories for smoke bombs, gas pellets, and even a homing device no larger than a marble. And the pentagram tie clip was a diamond cutter, while the ankh lapel emblem opened out into a blasting device which could penetrate two inches of steel if it had to.

But that wasn't all – crusaders always have more.

Philip St George, eyes narrowed in concentration, testing the velvet bonds by flexing his corded muscles, suddenly recalled another item of the highly technological arsenal which St George millions and Sidney Kite know-how had arranged to be made and delivered to him when he had begun this incredible crusade. Arming himself to match a platoon.

He rolled over on his side on the cool slab of table, facing the left wall, doubling his legs behind him so that his fingers could reach the heels of both uplifted feet. Muscles rippled on his body.

The tips of his strong fingers, pliant, talon-hard, dug into those heels. Probing, digging, removing the minute, flesh-colored strips of adhesive tape. Under each layer of concealing flap was a one-inch length of something. Something that glittered in the eerie light of the foyer. Something sharp as a razor, no more than one eighth of an inch in diameter. Something that could sever anything less than an iron bar. Compressing his lips with the effort of reaching these objects, holding his breath so as not to gasp aloud and signify that the prisoner had awakened, Philip St George maneuvered the tiny instruments of release very carefully between the balls of his forefingers and thumbs. His eyes shone with a fierce, cold glare.

Which was when the second factor entered the scheme of things.

The accidental one.

The gift.

The unexpected boon. If that was what it was to be. . . .

Poe, the black cat, ambled into the strange room. As onyx and furry and complacently silent and undisturbed as ever. He mewed faintly and the sound made Philip St George's head come about in quick alarm. Poe's green eyes were unblinking as he watched the threshing, bronzed, naked man on the stone table. A strange sight.

He moved further into the room, gliding like a shadow.

Philip St George watched him, unblinkingly too.

His heart had almost stopped beating. His blood iced over.

He could not know what the cat would do. Cats were, at best, oddities.

Trying not to panic, keeping his nerve, Philip St George kept on slicing away at the velvet bonds on his ankles, never taking his eyes off Poe. He had no superstitions about cats. Only wariness.

And a healthy respect for their unpredictability.

The moaning, murmuring chorus of the voices somewhere close by seemed to raise in volume. As if the owners of the voices were drawing nearer, approaching this odd room that held him captive.

They were coming for him, in all probability. With all their infernal hoopla, hokum, and devilish games. Rigged up and decked out in the paraphernalia of their ugly occult rituals.

The coven, damn them.

Sister Sorrow's coven.

What else?

The same people who had been responsible for the slaughter-house deaths of five helpless females. Including the cyanide finish of poor Norma Carlson. *Christ, he had to get free . . . now! Now!*

Poe suddenly hissed sharply, a shattering whistle of sound.

Philip St George twisted frantically, sawing at the velvet ropes, gaze still pinned to the crouching, arched shadow of an enormous black cat stalking him from the corner of the weird foyer.

And then there was nothing left to do but fight back.

For Poe drew himself up into a balled, clawing mass of ebony massiveness and shot from the floor. Rising like a tiger, green eyes glittering, face contorted in rage. All at once, with dazzling violence and a lightning burst of feline speed, Poe was airborne.

Shooting toward Philip St George with fury burning in each eye.

With a blur of dazzling swiftness.

A liquid spurt of jet and terrifying animal determination.

Toward Philip St George's defenseless face.

FIRE UP ABOVE

The Satan Sleuth sawed with almost insensate speed.

The velvet bonds which lashed his ankles together came away just as it seemed there was no time left at all. Not a second.

Poe's powerful cat body, deceptively soft and smooth in appearance, landed on Philip St George with all the force of a released projectile. Fortunately, and with that combined mental and physical coordination which had remained with him all his life, the naked giant was ready for Sister Sorrow's familiar. On the mark.

Racing time and flying motion unified in an almost indescribable sequence of events. Whatever else Poe was, he could not have expected what now happened. Feline fury transformed to immediate grief. Philip St George, legs free and loose for action, was an opponent of a far different stripe. Poe never knew what hit him.

His spitting, fanged face, primed to bite into the defenseless bronzed face before him, suddenly encountered a stone wall. Philip St George's skull spun down. Poe slammed into his cranium. St George had averted his face at the last possible instant, swiftly. Squalling, hissing, Poe's body rebounded and hit the floor in a sliding sprawl. In a second he recovered his balance, but now the stranger in his mistress's house was fully prepared for him, even though his wrists were still chained together with velvet thongs. Yet Poe blindly, now thoroughly outmatched and foolhardy, plunged back into the fray. Just as the voices close by rose in an almost feverish pitch of what sounded like some expression of religious excitement. The din, more Greek chorus than ever, filled the walls of the foyer.

Poe came hurtling, scratching, clawing, spitting his rage.

Philip St George, who had matched wits with Bengal tigers and charging rhinos, did not make the mistake of underestimating this smaller member of the cat family. The dimness of the vestibule, with its confounded stone slab bed-table, left little space for advantage. Or error.

Still Poe was no more than a cat. A household creature, whatever his damned station might be in Sister Sorrow's occult scheme of things. Philip St George had to silence him or all was indeed lost, forever.

He did. Neither with regret or mercy. Just cold-blooded execution.

With an accurate, unerring, jet-propelled thrust of his right leg. His bared foot, extended and leveled in the best Karate attack position, might have been composed of stone, pure bone, or solid steel. The effect was the same. Poe, on the explosive fly, was caught precisely where his furry neck met his green-eyed, whiskered face. There was an awful splat, a muted crack of something breaking, and the black, unleashed body shot to the wall, struck with thudding force and impact, then slowly dropped to the darkened floor. Poe didn't have time to let out so much as a whimper or a mewing noise. He was now no more than an ebony blot on the parquet surface of the foyer. Philip St George did not pause or hesitate a second longer. Time was now the very most important single item left in the whole wide world. The Dark one and the Light one. The commotion of the struggle, as brief as it had been, the speeded-up passage of the interval during which he had been left alone, had made him lose all track of the hour. Urgency and that normal amount of cold fear which made him a human being as any other now served to make him a no-lost-motion whirlwind of endeavor. Spurring him on, firing his purpose, were the uplifted, fervent voices chanting, chanting . . . seeming only on the other side of the wall to the left of the strange door which was Sister Sorrow's boudoir. He didn't know that, of course, but he found out soon enough.

For he razored the velvet ropes binding his wrists, gathered up all of his Peter Bond masquerade, filling his arms, and quickly quit the foyer, going through the door facing him. He didn't give Poe the black cat a second thought. Though killing animals was no longer his game, especially what was purported to be a domesticated cat, there had been no choice. No alternative. It was the difference between life and death. He had chosen life, as he always would. Just as he had made the same decision when faced with the gruesome destruction of the woman he had loved up there at Lake Placid. He could have committed suicide then – he'd had all the romantic reasons, the desperately empty motives, but he hadn't killed himself. He had gone on.

Poe he could not grieve for. Not for Poe, a killer.

Not a Dark destroyer who could have cost him his life.

There was no more room to think, to dwell on that.

So he rushed into whatever safety lay in the area behind the dark door. Not knowing what he was getting into, where it led, or what it might mean in terms of his safety in Sister Sorrow's Manhattan eyrie.

It was like walking into illuminated nightmare. Into a crazy house.

Even as he snapped the door latch behind him, securing himself against the frenzied medley of chanting voices which still pervaded the atmosphere, even in here he could scarcely credit his eyes.

Or his senses. It was like a trip sponsored by LSD intake.

He swore under his breath, moving forward, placing the rumpled pile of Bond's identity and equipment on a squat, low black marble table just to the right of the door. For a long, chaotic moment he compelled himself to think, to make a mental adjustment, a visual orientation of the universe of mystic nightmare surrounding him. Alarm bells tingled in his brain. His muscle tensed, his pulse quickened. The base of his spinal column experienced that cliché but nonetheless true sensation of cold, clammy dread. The graveyard touch. The mortuary chill. The cemetery iciness. The room, if room it was, stood starkly incredible, other-worldly and totally unreal. Ignorant though he was of the layout of the penthouse complex of coven, ghouls, and sorcerers, he instinctively knew where he was.

This was it. It had to be.

The very heart of the house. The core, the nucleus, the soul. The nerve-center of a witch's domain.

This must be the place.

Where all the madness, the mysticism, the mumbo-jumbo and quackery were born. Modified and perfected. Sanctified and made rational and feasible in a civilization of modern men and women who should have known better, who should have laughed and walked away from such blatant and obvious chicanery no matter how effectively and dazzlingly executed. Like a stage show designed by experts.

Philip St George saw and understood all that in one glorious burst of reason and common sense. And ratiocination.

He moved deeper into the weird boudoir. The bizarre chamber.

He saw it all now.

When it counted the most and yet perhaps when he was in the worst possible position to do anything about it.

For he himself was caught, fly in the spider's web, to all intents and purposes, unless he could whip up some choice witchcraft of his own. The coven was gathered – and he was all too aware of that – for that damned, repetitive chanting of some infernal litany was still bombarding his eardrums. Almost louder, too. They were closer than ever. '... *Domine, Satanas ... Omni ... Omni ... Omni ...!'*

Philip St George collected what was left of his mental and physical resources and studied the room. Appraising all it held and containing his own wonder, uneasiness, and dread. Something had to be done.

He saw ... and barely dared to believe what he saw.

Not a bedroom. But a witch's kitchen. Colonial, Salem-style, brown wood, oaken, raftered, and seeming like a page out of the ancient past. With terrible, magical, no, Satanic changes in design and mood.

There was the red-brick hearth fireplace, complete with Dutch oven, small bellows, cast-iron hanging rod, roasting spit supported by andirons – wicker, wooden chairs, fowling piece on the wall above the hearth, dangling ears of corn, long wooden shovel – all proper and truly a facsimile. Yet charred, smoldering coals burned in the fireplace, sending flickering, hellish flames licking upward, flinging ghastly silhouettes over the floorboards, scatter rugs, and furnishings. *But the room was cold.*

And that was the least of it.

The rest of that devilish apartment more than fulfilled the awesome promise of a woman of Catharine Copely's witch-like beauty and wickedness.

Philip St George also saw . . . The Impossible. The *It-Can't-Be!*

The black bats hanging suspended from the central beam overhead. A horde. Their fiery eyes glowed in the semidarkness, their bunched, unmistakable outlines sending flurries of iciness up and down his back.

The square, cross-legged wooden table with its surface crowded with every conceivable shape and size bottle and container. Each item was labeled with cabalistic signs and emblems, signifying contents whose nature God only knew. The air of the room was charged, incense-laden. Like the roasting of chestnuts blending with the stench of decayed carrion.

And there were other things . . . all implementing the madness.

A huge yellow pentagram with the Satan sword fixed in the dead center. The firelight made the painted circle dance strangely.

There was an ax, too. A short-hafted chopping ax.

Shining, scrubbed clean so that it gleamed like silver. Like fire.

The ax lay harmlessly on a smaller wooden table just before the brick hearth fireplace. But the vicious cutting edge sparkled.

And there were crescent-shaped moons pinned in an almost irregular pattern all about the four walls of the room. Showing like obscene graffiti.

And ouija boards, planchet boards, prayer wheels, crystal balls. Greek symbols, Egyptian heiroglyphics, a bed of nails, an Iron Maiden, a miniature guillotine . . . indeed, the whole

9

catalog of artifacts which a scheming, beautiful woman might well employ to keep a coven in line.

But more than that, even incredibly more incongruous and deadly than the small, antique bed placed in one corner of the room, set back from the motif of colonial kitchen, was the last remarkable sight presented to Philip St George by the make-up of that chamber.

It wasn't a page from history, it was a page from Hell.

A descent into the pit of people's minds, follies, and cruelties.

The topmost totem of Evil. The last barbarism.

The hanging bats, who had not moved but whose vivid eyes continued to gleam without blinking, were one thing. Loathesome though they were.

This last was something else.

He hadn't spotted it right away because there was so much to see, to take in and digest, all in a glance. All in precious, valuable moments.

But now he saw. And shuddered.

There was a long row of hellish masks on the wall of an alcove set back from the rest of the room. Satanic representations, with horns, frozen leers, curling, spiked mustaches and beards. Lifelike to an amazing degree but not even that stunning miracle of the wax-museum art was what made Philip St George fight to control his emotional reactions.

Below the masks, of which there were at least a dozen, was a mass of something. A huddled, shapeless, small bundle of some kind.

He drew closer because he had to, passing directly beneath the overhanging squad of black bats, but he ignored them. They did not move.

There was a woman in the alcove.

As small as a child, naked, and – unconscious.

He did not see the incredible floor-to-ceiling mirror at all. Not at first. He only saw the woman. And rushed to her side.

She was lying contorted, knees nearly up to her chin, arms fanned out in a gesture so pathetic that she seemed crucified, if not dead. Yet that in itself was not what made Philip St George's mind turn over in revulsion. And white-hot fury.

Someone had used the woman's body for a pin cushion.

She was bleeding from at least a thousand cuts.

And all the cuts were cabalistic symbols, Devil's markings, cult handiwork – the most outstanding and vicious of which was the jagged, irregular, large pentagram star which extended from the crest of her small mounds of bloody breasts all the way down to the rim of her pelvic cage. The woman, girl was more accurate, was literally a mess. Something for *Le Messe Noir*, too? *The bastards!*

Philip St George didn't know. The mirror was just behind him.

But even as he stood, trembling and temporarily impotent with blind rage, in that insane room which was the heart of the demoniac den he had invaded, he swore that he would find out. Find out or die.

And that damned irreverent, fiendish chanting seemed to grow louder still. Louder and louder . . . as if the walls held the voices.

'. . . *Satanas . . . Satanas . . . Omni . . .* '

The coven was not in full swing, really. Not at all.

Sister Sorrow was merely leading up to the startling surprise she was about to spring on Brother Nicholas, alias Duane Farmer.

The Black Mass would not begin until Brother Nicholas learned exactly where he stood in the new order of things. No longer tall.

When Sister Sorrow would select a new celebrant. A replacement.

Who would have to be awakened and made ready, whether he cared for the arrangement or not. Sister Sorrow was sure she could persuade the bogus Peter Bond to join her in unholy matrimony rather than go down the black primrose path as victim and offering. Better to be mate to the Divine Sorrow than be sacrifice. Love rather than blood.

That made sense to a beautiful witch.

It was something Philip St George didn't know, either.

But there were a few things Sister Sorrow did not know herself.

A few surprises, in fact.

Philip St George quickly set about arranging them.

The first thing he had to do was find out if there was any-thing like a telephone in Sister Sorrow's own version of Night-mare Alley.

There was nothing he could do for the naked girl just yet.

She had been drugged beyond belief. The main line-treatment – the arm.

It would be hours before she awakened . . .

If that damned Devil's chorus didn't do the job first.

They were coming out of the walls now.

As if they had the weird kitchen-bedroom-torture-chamber boxed off on all sides – like the arena at the Coliseum.

Christians, five minutes!

Despite the terror all around him, Philip St George's tense and handsome face froze in a taut, death's-head smile.

Gallows humor always had its advantages.

No matter what happened next.

Arthur Fling fairly leaped at the telephone when its sudden, strident clamor upset the peaceful stillness of his office. The long, snow-filled day, the lack of any further developments in the cult murders, as well as no fresh contact from his movie-star-imitation crank caller, had left him rather high and dry. A condition not dispelled by young Terry Lenning's chortling, gratified agreement with the crank caller's recorded opinions of just what five pretty girls without heads found in strange lonely places was all about. Fling had given in, albeit grudg-ingly, phoned Police Headquarters, and shared the facts and details with them. Voice-Print had come up with nothing; there was no suspect in custody to compare the movie-star imitations with, anyway. Yet Lenning had been busy anyhow. Though the FBI and the police had still made no connection with the strange poisoning of Norma Carlson and the ritual murders, as the case was now on file and known as, both at Centre Street and the Manhattan branch office of the Bureau, Lenning had suddenly noted the similarity in size and general physical make-up of the five corpses whose wills were still being probated and whose cadavers were yet to be properly buried, pending official solution of their murders. There were no relatives or angry lawyers to press the issue, it seemed. But

Lenning was adamant about one single aspect of the entire string of homicides.

A Satan cult was at work. Motive, beyond the usual the Devil-made-me-do-it cop-out, and yet to be determined and established.

Lenning flew to the voice-recorder and a matching phone as Fling quickly semaphored that the sudden call was from none other than Mr Boris Karloff himself. The strange informer with theories of his own. Reasons – like the cult – still to be discovered.

It was very late in the cold, dismal day but both agents had always been the sort of men who remained on the job long after office hours. Neither was married but that had little to do with it. The job came first with agents like Arthur Fling and Terry Lenning.

A taping was made anyway, as was the custom. The office routine.

Though this call was far briefer, and as was noted later on, in the light of day, when time was not so important anymore, the Karloff imitation was hurried, breathless, and held utter urgency and a rather frightening sense of no-fooling!

Voice: 'Mr Fling, it's me again. The mimic informant.'

Chief: 'Oh, it's you, my movie friend.'

Voice: 'Karloff speaking. But skip everything and listen to me. I'm only going to say this once. There is no time for questions and answers. I have your witch for you. And her coven. And you must come now! With all possible speed. Bring an ambulance or a doctor. Or both. There's a Black Mass going on right this minute – '

Chief: 'Hold on. You're going too fast.'

Voice: 'Shut up. I can't speak much longer. Write this down. Don't trust your memory. Miss Catharine Copely, One Sixty Central Park South. Penthouse. Come! Come now. Explanations will have to be later. I'm hanging up now.'

Chief: 'Copely. Central Park South. Penthouse. Check. But – '

Voice: 'Hurry, man, hurry! For God's sake . . . '

That's all there was to the conversation. It was rapidly terminated.

There was a gasp, as if the caller had suddenly been interrupted, a sharp click of the unseen phone being replaced on its cradle. Fling shot a quick look at Terry Lenning, drummed his right hand on his tidy desk, looked at the round clock on the back wall, and then nodded, as if to himself. When you're a chief agent, you take chances, act on gut reactions, and take the initiative. Even if you risk playing the fool.

Lenning was already tightening his tie, face expectant, spoiling for action. His astonishment with the caller still showed, however.

'We take a chance, Art? Or do we do it by the numbers?'

'We go for broke,' Arthur Fling snapped. 'Two squad cars, the full treatment. Riot guns, warrants, tear gas. Call Homicide. They ought to provide the ambulance and it strikes me they have a right to be ringside for this. If we're making saps of ourselves . . . ' His voice trailed off as he shuddered. Terry Lenning grinned and reached for the phone on his desk, moving expertly and smoothly.

'Fling, I love you. And the laugh would be on me too.'

'Thanks for the thought,' Fling said bleakly. 'Some helluva consolation. It'll be my neck, not yours, Terry me lad.'

'I trust Boris Karloff.' Lenning laughed but he meant it. 'If he found a witch on Central Park South – ' His call was connected and he broke off, barking into the transmitter. Fling was already in his gray topper and large borsalino. He had strapped on his issue revolver. Something he hadn't used in years. Not since this desk job.

'Come on,' he urged Lenning. 'Let's get cracking. The man sounded like he was in trouble. He was talking plenty fast.'

'Wonder why he didn't call the cops in the first place?'

Lenning was alternately cupping the transmitter and talking to both his superior and the people down at headquarters.

'They get thousands of crank calls daily, Terry. Probably didn't want to risk being lumped with all the kooks. Can't blame him. Besides, he's got what he wanted. Here we are rushing off into the snow, dragging Homicide along for the ride.'

'Let's hope it's not a real ride.' Lenning hung up and started hitting buttons on the intercom box which connected with the offices of other agents in the building. 'This is the first break in this whole cockeyed slaughterhouse. It's gotta be on the level.'

'It better be,' Arthur Fling said grimly. 'Or I'll be getting a shine on my britches behind a desk in Split Lip, Nevada.'

'Better Split Lip than Outer Mongolia,' Terry Lenning answered with a smile. But the manhunter look had not left his alert young eyes.

The crank caller made sense to him. A lot of sense.

If he were in danger when he called, and still wanted to mask his voice, what better voice than Karloff? That soft, gently English speak-box with its rises and falls. No crank could be so clever.

And anyway, Lenning still shared all of the informant's theories and notions about the strange slayings. About Satan worship, about the whole ball of wax. The Devil was in New York, as far as Lenning was concerned. And if he and old Art Fling could crack the case –

Funny about the victims all being so small though.

Cult or not, it wouldn't have taken too much strength or power for *one* person to have handled that kind of a kill. Especially with an ax.

One tough, powerful person might have been enough.

Man or woman – even a very strong teenager . . .

There was no more time to think about that though. Not now.

Chief Agent Arthur Fling was hurrying him and a group of quiet, grim-faced Federal agents into waiting cars, getting the show on the road.

And the damn snow was still falling, mixed with rain and soot and pollution and foul city emissions. The town never looked uglier.

It was a great night for a murder, as the old cop saying went.

And an even better night for a witch hunt.

On Central Park South.

Far uptown, safe and warm in his own apartment on River-

135

side Drive, Mr Sidney Kite could have been forgiven if he had felt the same way. The bitter end of another long, exasperating day at the office had still brought no word from young Phil St George. But Sidney Kite was a genuine old dog who could not be taught new tricks.

He kept up a constant barrage of phone calls to the Explorers' Club and the St George apartment. To hell with leaving cockamamie messages with strangers on the line, he just might catch Phil at the right time. In a mood to do some talking.

In spite of all he had said to Miss Irene Walters, Kite *did* love Philip St George exactly as a man loves the son he does not have. And might never have at the rate he was going.

The radio blaring in Kite's living room was small consolation. Along with the rotten weather, an airliner was down in the Rockies, Dow Jones had taken another terrible beating, the price of paper was skyrocketing, and Mr Nixon's reputation was getting worse than ever. The Kite ulcer rumbled warningly so the lawyer flicked the radio off in disgust and made himself a glass of warm milk.

And with that act missed a very important news bulletin.

A flash report containing some vague details about a mysterious explosion and fire on Central Park South. A four-alarmer.

The announcer indicated that there would be more to come.

As eyewitness reports came in.

THE VERY DEVIL

The coven had indeed gathered.

In the vaulted chamber with the darkened walls, the *Sigil of Baphomet* glowing starkly by wavering candlelight.

All of the hooded conclave, spear-headed and driven by the tall, amazing woman they all revered as Sister Sorrow, had experienced the shock of their cult existence. Their collective, distorted personalities and minds had been witness to living fantasy. Fantasy which was to expand to vivid, unspeakable proportions before this night of nights would come to an end.

Brother Nicholas, Duane Farmer by daylight, felt the shock more than anyone present. Its tremors and repercussions would be fatal for him. For Sister Sorrow had at last truly shown her fine Satanic hand. As soft and slender and shapely as it was.

The manner in which she figuratively turned the tables on her co-conspirator, Brother Nicholas, transforming his high status in the heirarchy of the coven to that of lowliest outcast, was sudden, startling, and overwhelmingly cruel. Savage, too.

A turnabout of Devilish design, monstrous form – yet completely in keeping with the Devil's own code of fair and foul play. Winning was all that mattered; the end justified the means and only the victor should have the spoils. Lucifer would have laughed.

It had all begun so seemingly harmlessly, routinely.

The congregation as before, the circle taking places.

Sister Sorrow had led the Dark brothers and sisters in an invocation to Satanas. With Brother Nicholas at her regal side, gaunt and characteristically somber in his celebrant's cowl and cloak.

As the fervid voices rose in a frenzied Messaniac chorus of love and devotion to the Master of Darkness, the Divine Sister made her first move. With typical, velvety tones of cultured emphasis.

Brother Nicholas, in effect, never knew what hit him.

He had been far too busy distributing the malevolent equipment for the pending *Le Messe Noir*.

The fiendish artifacts, so relevant and important to the forthcoming procedures, had been doled out to the proper parties. The incense, the candles, the gong, the crucifix, the Bible of Satan, the mock wafers which profaned the Christian Mass – all of those very necessary and proper instruments and devices were in the hands of the designated role players. The ceremony was all set. Armed for action.

Brothers Herod, Attila, de Sade, Mars, Hitler, and Judas were primed and ready. Impatient to get on with the ritual. Sisters Lilith, Bertha, Circe, and Medusa all trembled feverishly with shuddery anticipation. Sister Sorrow had whetted all their appetites with her evil announcement the night before of something unique and truly classic in Black Masses. No one could contain his anxiety and expectation.

The zealous chanting, the throbbing harmony, the flickering shadows, the blazing candles, the invocation warm-up, had only served to heighten everyone's sense of desired exaltation. And delightful wickedness! Something the Mass had always provided – *Ave, Satanas!*

When the hymn of fealty at last was done, all would file into the darker chamber which the coven employed on days of inclement weather. The December snows had not abated this day. The coven therefore would remain indoors. Sister Sorrow was glad of that. No end.

For what she had planned, and was about to do, the interior of her domicile was far, far better. The circumstance of foul weather had been a gift from Lucifer Himself! An emperor's boon of love.

Betrayal and the act of replacement did not need the light of the full moon. It would be much finer and properly ritualistic by candlelight. In red and yellow flame. As befitted Devil's rites.

Yes, it would be much better this way.

This moment.

This time.

Now.

Suddenly she raised one eloquent arm, slicing through the voices blending in harmony, halting the chanting of Latin prayer.

The voices all stopped, as if a wall switch had been pressed.

Hooded figures shifted uneasily. The chamber rustled with small, indistinct noises. Brother Nicholas turned to the Daughter of Satanas. His cowled face showed briefly in a flash of surprise. Of doubt.

And perhaps only curiosity.

Sister Sorrow had never interrupted a devotional hymn before. *Never!* Not in his memory. Why – it was almost sacreligious! Profane by the Devil's rights. His gaunt figure moved closer to Sister Sorrow. But still he kept his distance. The required three paces.

'Who interrupts the words of love and loyalty?' he mocked flatly, without too much reproach in his voice for he did not dare that.

'I do.' Sister Sorrow's richly timbered voice rose on a note of anger and shame. 'There are ears here not fit to hear our prayers. Eyes not fit to look upon the *Sigil of Baphomet*. A body not worthy of the sacred robes it wears in the sight of us all. I, Sister Sorrow, have found a serpent in our midst!'

The coven gasped, as one, in a unified blurt of sudden wonder, amazement, and fear. Hooded heads swiveled, shrouded faces peered out, each devotee trying to see what Sister Sorrow said she saw, looking for the blasphemy in the face of a brother or a sister. The high priestess's words echoed like a death sentence in that darkened chamber.

Candlelight flamed over the tableau. Whispers rolled around the room. The inverted cross, high on the wall behind the brother known as Nicholas and the sister ordained as Sorrow, was now more arcane and unearthly than it had ever been. Someone was sobbing, low.

'Who?' Brother Nicholas growled. 'Who would dare such a treason on us all? Who could – '

'*Thou.*' Sister Sorrow's accusation knifed across his words. Without hesitation. 'Thou has been our brother, Nicholas to us all, but thou should have been named *Judas*!'

If Poe the black cat had still been alive to walk suddenly into that room, everyone present would have heard the tread of his feet on the parquet floor. The deathly stillness was nerve-racking. A blazing candle flared just then on its wrought-iron base and the spit of sound could have been a pistol shot. It was as unexpected.

Duane Farmer struggled to speak.

He could not.

He took a step backward, instead.

His cowled head wagged from side to side, as if the awful charge was too enormously absurd to be answered. The thrust of his tremendous nose, seen now in profile, never seemed more like the butt grip of an old Western shooting iron. Or more grotesque.

Sister Sorrow seized on his mute, stunned behavior to reinforce her scheme. What she was now about to do. What she had planned ever since she had made up her mind staring down at the unconscious bronzed god on the stone table in the foyer. Manna from Hades!

'See?' she challenged the watching and listening coven before them. 'See and comprehend. Brother Nicholas had been rendered speechless by his guilt. He cannot answer. But I will answer for him – and in the sacred name of He who leads us all to Divine Darkness – I condemn this brother who sullies our circle with his presence!'

The hoods and robes again shifted. The massed conclave, huddled like children now, seemed torn with mixed emotions. Someone dared to cough, another muttered, a third spoke in a low undertone. Brother Nicholas, still stunned, raised his arms in wild protest. Shaking.

'No!' he shouted, his hoarse yell nearly a scream. 'What have I done? What is my crime? I who have served you all and this woman – ' He was appealing to the coven now, his movements hurried and erratic.

Sister Sorrow leveled her burning eyes at Brother Nicholas. 'It is enough that I know you no longer believe. It is enough that you are a spy, sent here by the outside world to learn our secrets and destroy us! It is my knowledge of that perfidy that will condemn you before this holy coven. I will exact the full penalty. I am the law here – and it is I who will be your executioner. Who among you will defy Sorrow, the Divine Sister, Daughter of He who is our Lord Eternal?'

No one would, of course. The coven had but one mistress.

Or could, even if they had the nerve to do so. Not this coven.

They had not the courage or the strength. The Divine Sister's rule of the coven was total and absolute. Nobody would question her fiats or decrees or – whims. Brother Nicholas knew that better than anyone. He who had been in league with her from the very beginning knew well the hold she had over this pathetic cast of characters.

There was only one thing he could do if he wanted to save his neck. So he did it, with a speed born of utter desperation.

With a cry of anguish, tugging his cowl away from his huge-nosed undertaker's face, he whirled like an activated top. Away from Sister Sorrow.

To run, to flee.

To escape.

To get as far as he could from Sister Sorrow and the coven.

It would have been useless to try to defend himself while this crazy bunch was hopelessly locked in Sister Sorrow's hypnotic spell.

As the coven roared its surprise and disbelief, the tall figure of Brother Nicholas broke from the raised dais which he had shared with the sister and plunged toward the door on the left, beyond the massed circle of hooded cultists.

No one tried to stop him. Traitors have no friends.

They knew that would not be necessary.

Sister Sorrow could do what she wanted, the same way she had changed Brother Carmody into a green garden snake.

It wasn't necessary, after all.

Sister Sorrow raised her right hand, her magnetic eyes gleaming weirdly – and pointed a long, shapely forefinger at Brother Nicholas's desperately scrambling body. Just as he reached the dim, dark door.

What followed was the sort of electrifying display of witch's power and sorcery which awed onlookers, humbled reasoning intellects, stupefied low IQ's, and terrified skeptics. Paralyzing all logic.

Brother Nicholas stopped dead in his flying tracks.

As if he had suddenly encountered a stone wall.

A terrible cry ripped from his throat.

He teetered on his toes, twitched as if about to go into some kind of dance, leaping into the air as if performing an intricate ballet step. And then came down, heavily. Like a man having a heart attack.

With a sprawling, sickening, lifeless thump of sound.

The coven looked on. Blinking, horrified, unable to speak or move. Or think. The witchery, the woman, was beyond comprehension.

Sister Sorrow laughed. Her low, vicious, mocking laugh.

'He is not dead,' she intoned regally once more. 'That honor will come to him later. For it is he who will be our sacrifice. Our offering to Satanas. Tonight will be a glorious one for us all. We will welcome a new celebrant. A divine apostle. That is the true blessing I bring to you, my brothers and sisters. A newer, far greater brother! *Brother Adam.* The original sinner, the father to us all. And henceforth I will become *Sister Eve.* His mate. I promise you, O Coven! – Satanas himself would smile at this union.'

Whatever Lucifer the Prince of Darkness might have said or done about Sister Sorrow swapping celebrants in mid-Mass was never to be known.

For the next spin of the mad wheel of Fate was to be the weirdest turn of all.

In fact, Sister Sorrow was about to receive the greatest single shock of her own fantastic, unholy career.

True witch or not, it was – *hellish.*

As the coven stood in petrified horror at the amazing and rapid reversal of the order of things in the circle, and Sister Sorrow folded her arms, smiling evilly at the assemblage – *it happened.*

There was a burst of flaming, explosive smoke. A blast of noise, a rush of violent air, and the candles on their bases winked out. All at once. Fiery, crimson light shot upward,

142

geysering. In the vicinity of the dark door and Brother Nicholas's sprawled body. Exactly *there*.

A rising, puffing, coiling, scarlet surge of smoke, heat, and sound. The coven, to a man and woman, stopped breathing. Sister Sorrow might have been made out of marble on the raised dais. She was frozen.

The darkness, pierced now with lambent flame, distorted shadows of many other hues and shades, flickering, undulating, was suddenly brilliantly illuminated. As if rockets had gone off, pinwheels had gone mad in the riot of color. A mammoth haze of unearthly light spiraled, wavered, and then – held. Holding one spot. Where the door was.

Where now – a figure stood.

Exposed, revealed, magnificently presented.

Stood?

No, towered, blossomed, rising magnificently, magically, appearing taller than any man alive. Broad of shoulder, long of limb. A figure as monumental as any statue in all the museums of the world. A colossus of a man. Of a being – of a god!

Caped, all in gleaming black.

Horned, spike-mustached and bearded, bold eyes commanding and furious, powerful form ready to do battle. To lead armies.

With arms akimbo, the fantastic apparition glared across the dark room at Sister Sorrow, the flaming lights playing on his saturnine, unforgettable face. Glared and stood his fiery ground, unflinchingly terrible, uncompromisingly wicked and utterly uncanny in the weird swirling red smoke and coruscating lights.

There was no mistaking this man, this intruder.

Not for the coven, which fell to its knees, terrified, frenzied, struck dumb with awe and horror in the sight of supreme evil.

Not for Sister Sorrow, either.

She perhaps least of all. For the moment, anyway.

It was Lucifer who stood on the flaming threshold of the chamber!

The Prince of Darkness.

Satanas!

The Beloved One.

143

He had come at last to Sister Sorrow's coven!

The wrong world rocked; the empty universe moved on its invisible axis; hearts, minds, and souls collided in that awful room buried in the complex of the apartment on Central Park South.

Evil stars shone in the atmosphere, malignancy twinkled.

Lucifer was here.

The Devil Himself!

THE RED MASS

No one in that incredible room doubted that they were face to face with the Devil. No one – except perhaps the woman who had invoked his spirit and the dread of his name to hold power and control over the lives of so many other people. Catharine Copely, Sister Sorrow of her own private Devil-worship cult, for a long, supercharged, terror-filled moment believed in her own sorcery. Her own witchcraft. She, along with all her hooded colleagues, was incapable of movement. Of clear thinking. Of any worthwhile thought. *Satanas was here!*

The moment held. A flaming, smoking, nearly nightmarish single space of time, in which all the normal functions of human life seemed to be checked, suspended in space.

Which was all the supernatural being standing in the hellish half light needed.

Lucifer raised both arms, the black cape spilling beyond his enormous shoulders. Two powerful arms, rippling under coal-black outer garment, lifted outward, the bronzed fists clenched. As if calling for strength and determination, as if summoning all the willpower and devotion the cult might hold for him. All of the hooded figures before him, strung out in various postures of genuflection and fealty, heads bowed, bodies shaking, could hardly bear to stare up into that saturnine, Satanish face.

Sister Sorrow stirred on the dais, brushing at her eyes as if something were in them. The haze was eerie, still swirling in spurts and trailing vapors, comparable to all the colors of the spectrum. There was an unholy illumination in the chamber now. The very air was saturated with incense, the perfume of

145

evil. The ambience of glory and sin. And ineffable damnation.

The terrifying figure threw back its head and laughed.

A laugh like no other.

Mocking, strident, pealing, shattering the senses. A laugh from the depths of Hell.

The coven trembled en masse. Sister Sorrow shook her lovely head. Striving for clear thought, for an answer for something that *just could not be!* Whether she devoutly desired it or not.

Lucifer's fists opened, the ten fingers jutting out like spear-tips. Perhaps everyone in the chamber saw the small, round, marblelike balls which suddenly fell from those hands. Fell in a rolling, clattering, disorganized scramble across the parquet floor. If they had, it wouldn't have mattered anyway.

Glass tinkled with a breaking noise.

If Sister Sorrow saw, she too like all the others, could not have been prepared for the magic the rolling, spilling balls accomplished. Lucifer's power, if there were any lingering doubts in the minds of those present, was mightier than ever.

No one saw anything, no one heard anything. Yet the chamber flamed and smoked with the weird lighting. The only sound was Satanas laughing with rich venom and the coven's murmuring, awed voices. But Satanas stopped laughing and the coven stopped talking.

And the magic was performed.

Executed.

With lightning swiftness.

One by one the hooded figures, agitated, as if suddenly attacked by some unknown force, tried to rise. To claw out the cloaked arms. To run, to hide as Brother Nicholas so futilely had only moments ago. But there was no escape from the Devil and his wizardry. The brothers, de Sade, Mars, Judas, Attila, and Hitler, all fell where they cowered. Toppling like tenpins in shapeless sprawls across the floor. Bertha, Lilith, Circe, and Medusa, colliding with each other in a terrified, crazed pack of threshing figures, followed them to the floor. Only Brother Herod nearly gained safety. But then he too was cut down. Invisibly, magically, pawing at his throat as if he could not breathe. He couldn't. The odorless, colorless gas contained in the small round marble-sized pellets was like some unseen power which strikes from Heaven or Hell. The coven littered the chamber, each in the same attitude of sprawling, stupifying unconsciousness.

On the dais, dazed, unthinking, Sister Sorrow stared across the chamber toward the being on the threshold. She had not breathed, subconsciously saving herself, if only briefly, from the fate which had befallen the others. But now she too could not escape.

The tall, incredible figure of Lucifer was advancing across the floor, literally stepping through the swirling crimson smoke, unscathed. Like a god walking through fire. Like –

Sister Sorrow suddenly understood.

Instinctively, with phantomlike agility, she whirled from the dais, leaping for the other side of the darkened room. And another secret door. The door which was but another entrance and exit of this strange chamber. The door which too was in reality a long floor-to-ceiling mirror on its reverse side. Identical with the one from which Lucifer had appeared in all his horned and carnal glory. There was no time to think of reasons, explanations. Not for Catharine Copely, who was now discarding her Sister Sorrow self-hypnosis and behaving like any mere mortal in terrified fear of her life.

But she had not reckoned with the fantasy of the Devil in the flesh. Whoever and whatever he was. Demon, myth, or mirage!

He reached her in a single bound, coming up like a giant bat and enveloping her. Closing around her, embracing her, and with startling determination carrying her through the door on the very impetus of her forward rush. Catharine Copely's flesh burned at the sudden touch of strong, fiery fingers. And then she felt herself flung. Hurled outward like something despicable. The chamber door slammed and she slid along the floor, hooded cowl dropping from her head. Her senses and boiling blood were tumulting. A dizzying wave of nausea and dread was overwhelming her. She fought it off, shaking her bewildered head, rising from her scraped knees.

She turned. And saw the Devil again.

With her chamber behind her, all its confusing smoke and noise and magic suddenly removed by the simple closing of a door, she fought to gain control of her brain and senses.

This room was no more than a cubicle. A squared affair without windows but pleasantly carpeted and furniture-less. It had merely served as a connecting area between the chamber she had just quit and the large room just off from the penthouse salon. But never – never in all her life had she expected that she would stand face to face with Lucifer in this room. Staring at him, riddled with fear and doubt, trying to relate to what happened. To make some coherent sense out of it all. Lucifer had appeared, the coven had seemingly dropped dead, she had not. And now she was alone. A prisoner or what – ?

Shuddering, she looked at Lucifer.

And suddenly it was all quite clear. All too lucid.

A cloud had passed over her eyes and mind. The effects, the tricks, the staging, all had been masterful. Worthy of a Lucifer indeed.

But this was not Satanas!

This was a remarkable man, a giant, standing before her, arms folded, no more than ten paces away, eyeing her with a detached and almost cold-blooded amusement. As if she was something pitiful and had not a chance in the world to escape. To gain power again. To make her own cosmically evil world right once more.

'That's it,' the Devil said in a strong, flat, masculine voice. 'Stand still. Don't run and you will not get hurt. I want to talk to you. We have about twenty minutes to clear this mess up. Before the authorities get here and put the blocks to this ugly setup of yours, Miss Copely. Don't worry about the coven. They aren't dead. Just gassed. And they'll be sleeping if off for at least an hour.'

Catharine Copely couldn't speak. Not just yet.

But now she had breathing time in which to see all the things she should have seen right away. Startling yet simple things.

She recognized one of the Devil masks from the row of such contraptions in her private chamber. She picked out the cape as one of her weatherproof articles of dress for the Manhattan streets. The black outfit covering the powerfully muscled body standing before her was no more than a pair of her leotards stretched to the breaking point. Yet all in all, taken with the lithe and gracefully magnetic figure of the man she had left unconscious on the stone bed in the foyer, it was small wonder that anyone who had seen him, under such conditions as trick lighting and smoke effects, would have believed him to be Satan Incarnate. It was the man who had called himself Peter Bond. . . .

The mate she had chosen. The replacement. *Great Beelzebub!*

'Take off your mask, Mr Bond,' Catharine Copely said dully, tasting the defeat in her mouth. 'Who are you really? What are you to come and spoil my private life like this?'

'Private life?' echoed the Lucifer mask, not obeying her request. 'It would have been perfectly all right if you had kept it that way. Thirteen oddballs catering to their Devil worship. You wouldn't even get arrested for that. But no, my not very dear lady. It was anything but private. You slaughtered six women to further the hold on your cult. Women who had what you really wanted. Their money, their lives. Six women who all must have signed some kind of will leaving everything they owned and had to you. I suppose you'll wait a year or so before you step forward to claim the money. Or is it all in the name of some foundation – the beneficiary? Society for the Brotherhood of Satan or some such? I suppose it would be. Even Norma Carlson, as bright as she was, had fallen under your spell. You chose your victims well. Lonely young women, without family, coming off a lost love or a sad love affair and welcoming your cult with open arms. There's a touch of lesbianism in a beauty like you, also. That would help in any circumstances with women whose defenses were down. Tell me, Miss Copely, how close am I to the truth? Remember, we've only got twenty minutes.'

Catharine Copely shuddered, hugging her arms in the folds of the cloak. There was still something so sinister, so awesome, about the height of the man before her, the frozen leer of that Satanic mask.

'How much do you really know?' she whispered.

'Enough.' The voice was smoothly casual, without emphasis or urgency, yet it was a commanding voice all the same. 'I know all about mechanical bats. I saw your hanging collection in the boudoir. Neat and effective. Glowing eyes, whirring

149

wings activated by a small motor, even a wire hook-up so that they can be maneuvered like a kite from the roof of a building down to a window. You were either warning me, I suppose, or just setting me up as a new prospect for your little group. Who handled that operation for you?'

'Brother Nicholas . . . the man I was going to replace with you. You see, when I undressed you . . . and removed your disguise . . . well – oh, whoever you are, don't be a fool! Share this all with me! The world could be ours . . . our own private world of money and power . . . '

She had lifted her arms, smiling now, turning on everything she had, and it was considerable. But Lucifer laughed and held out a long, pre-emptory arm. 'Hold it,' he mocked. 'Stay where you are. I saw how you treat the men in your life. Through the other side of that fancy looking-glass leading into the next room. How did you work that stunt? Or don't you want to tell me trade secrets?'

She shrugged and held out her right hand, tugging back the cloak that covered her arms. There was a strap harness rig fastened to the lovely wrist. A small tube was mounted on the rig. Parallel with the arm.

'Blow-gun device. You merely close your hand and a dart is released. A drug. Only a knock-out potion.

'Clever. Your coven really ate it up. You made them swallow everything you did. But there is one more thing I'd care to know. Tell me about Norma Carlson. And then you're going to write all this down and sign it. For the authorities. Before they get here.'

'The authorities?' Catharine Copely's exquisitely beautiful face contracted in a frown. 'Now, you're being ridiculous. You cannot prove anything.'

Lucifer laughed again. A short, hard laugh.

'You're forgetting something, Miss Copely. The pathetic little woman in your boudoir. All cut up and drugged. Did you kidnap her? Is she another rich young woman without a family leaving a will to you and yours? How will you explain her to the law? And remember – I can still kill you. And I will if I have to.'

150

'But why? Who are you? What do you care? Why should it matter to you what I do? What I am?'

It was Philip St George's cold blue eyes which now glared from behind the mask of the Devil. The eyes of the crusader and avenger.

'I hate the Devil. Didn't I tell you? What I hate even more is people who use him as a religion to prey on other people. I also tend to hate very much people who kill people I like. Norma Carlson was a friend of mine. I liked her.'

'Oh,' Catharine Copely said in a small, still voice.

'Oh, indeed. Now take it from there and answer my questions. I know why you killed her. She was ready to blow the whistle on you and somehow you found that out. She had dinner with Philip St George but that had nothing to do with it. You had already given her cyanide. Tell me how you did that. The authorities are still baffled, as they say in the daily newspapers.'

The woman who was Sister Sorrow now threw back her incredibly beautiful head and laughed. Derisively.

'Kill me. If you must. I have my secrets. I intend to keep them.'

'You can keep them forever if you are dead.'

'Would you really kill me? A woman as beautiful as I am? A woman who can give you so much. All this – this house, this cult. This power – we could rule the city.'

Satanas stared at her. It was unbelievable, really. He had told her, more than once, that the police were coming – if they did come – and she was behaving as if they were a figment of his imagination. A bluff? Was that it? Or was this strange woman before him just a trifle insane? Blinded by her own playing with magic and evil? It was difficult to tell. The line was too thin.

'You are weird, lady,' he said without anger. Almost pain. 'Who wielded the ax which beheaded those women? Tell me that, at least. There was no magic in any of that ghoulish stuff, was there?'

'No!' Catharine Copely snarled with visual relish, her green eyes sparkling. 'Not magic. But zeal, dedication, the fire of power! And it was I who chopped their heads off. They were my imps, my tiny girls, like children. Just like the one in the

151

other room. Mary Vernon will also die the same way. They come with me to the place I pick. A desolate place. I initiate them into my order by marking the *Sigil of Baphomet* on their white bodies and as they kneel before me, it is so simple to cut off their heads. The coven never knew but they understood. It was my hand that worked the magic. In that way, my control over them was absolute!'

'Tiny?' Lucifer echoed, suddenly understanding his own error about a collective cult massacre of a victim. 'So that's it. Very well. You'll write all that down, too. And even if you won't tell how you worked your "magic" on Norma Carlson, it won't matter. The Police Lab will solve it some day. You and your witch's brews.'

'I will sign and write nothing.' Catharine Copely drew herself erect haughtily. And now, oddly, the concealing cloak seemed to slip back, revealing marble, statuesque shoulders. Pale ivory skin glowed. Lucifer stood his ground, shaking his head. The spiked mustache twitched.

'That won't work, either, lady. You're not a woman to me. You're cold, heartless, unthinking, uncaring – a ghoul. A monster. Stop the act.'

She didn't hear him, obviously.

Or she didn't very much care.

Perhaps she was certain of the charms of her incredible beauty of face and body. For suddenly the cloak fell away. All the way to the floor. And abruptly she was naked. Splendidly, paganly naked. Like something only dreamed of, never seen in the flesh.

'Really, Lucifer?' she purred.

'Really.' His tone was flat, contemptuous, and mocking.

But she won anyway. In her own perverted, fiendishly clever way. She knew what she had been doing all the time.

As Philip St George discovered in a very few seconds. A few chaotic, never-to-be-forgotten seconds. Of doom and tragedy.

She was moving toward him, disobeying his injunction, green eyes luminous, red lips parted, all the rest of her tremendous physical assets on display. Undulating silkenly, maddeningly. The roseate blurs of her trembling breasts, the arc of her sculpted hips, the rhythm of her quivering abdomen, the magnificent fall of her long, dark hair, all of that – the Lucifer mask wagged almost sorrowfully.

And the Devil brought up his right hand.

In it a weapon shone.

The Greek ankh lapel emblem. The sign of life, which in this case was actually a blasting device of mind-boggling power. But he only meant to threaten the lady with it. Not make use of the armament.

But Catharine Copely had other plans.

All of which had to do with the voluminous cloak she had allowed to slip off her figure to the floor. In the process of doing that, the heavy garment easily concealed a button-device hidden under one corner of an oblong strip of linoleum tacked down beneath her feet.

Lucifer, who had spoiled most of Sister Sorrow's plans with some fantastic magic tricks, now found himself the recipient of one of the lady's fantastic assortment of sorcerer's wizardry. It was the final touch of Sister Sorrow's evil.

It wasn't only magical, it was downright inventive.

A pip of a performance.

With the lady coming on, smiling lasciviously, her nakedness like a direct blow to all the normal masculine instincts, the Central Park South complex registered one more trick of time and place. Catharine Copely had designed the eyrie with pure genius.

There was a rolling, rumbling, sliding noise.

Philip St George jumped back, eyes shooting, weapon up.

His last sight of Catharine Copely was her green lambent eyes.

Within an electrifying instant, she was gone.

From view. From before him. In an instant.

All because an impenetrable sheet of metal had slid down from the overhead ceiling, literally cutting the room in half, leaving her on one side and he on the other. He was too late to leap to her or avoid the metallic barrier coming down like a released guillotine.

It caught him a glancing blow before he fell back, against the wall, head ringing, dazed. But not unconscious. And then all he could see was himself in the barren half of the room and the

153

sound of her footfalls running away. Grimly, he pulled himself together, shrugging off the agony of his head, and lurched back into the darkened chamber. The sleeping coven scarcely stirred as he threaded his way past their recumbent figures. The fire-bombs and smokebombs had all dissipated now. Only the stench of burnt powder filled the air. And wax.

There was no more time to lose now.

He was sure of that when he heard the sirens keening in the December night, somewhere far below the penthouse floor. There was some satisfaction in that. The FBI had not ignored him, not dismissed him as a crank. When they burst into this setup, never mind the lack of conclusive evidence, Catharine Copely would have an awful lot of explaining to do. Everybody would, damn their black hearts.

The satan Sleuth wasted no more time. Or motion. It would be impossible to find Catharine Copely in this maze. Not worth it anymore, not with the authorities coming on in droves.

Even as he raced through the jumble of weird, unearthly rooms, he divested himself of the Satan mask and the cape. Quickly he flung himself into the Peter Bond clothes, without bothering to make up or even put on his shoes. The wailing sirens were louder now.

He gained the terraced garden patio with its Bonsai trees and foliage-shrouded mystery, found a connecting roof which led to the next building, navigated it with a tremendous ten-foot broad jump, and then vanished across the tar-slickened roof into darkness. The thinned-out snow, still falling with its overload of rain, pelted his flying figure. And the Satan Sleuth was gone. Once again.

Into the night which had brought him. Like a shadow.

Which was why he missed the holocaust that followed.

In a closeted corner of her penthouse complex, Catharine Copely was just as rapidly seeking egress from her crumpling world, the one which the fantastic stranger had briefly flung off plumb. But it would all work out anyway. She would see to that. Let the police ruin and spoil this coven for her. They had no real proof. Even the drugged Mary Vernon could be explained away, somehow. But first she must be gone. Not found here. She could always claim that the coven, friends of hers, had used the penthouse in her absence for their filthy sport.

And she might have gotten away with it, at that, if fate had not taken a hand. A fine, serendipitous hand.

Catharine Copely stumbled over the dead body of Poe in the foyer. That and the false-bottom shoes of Peter Bond, with their assortment of pellets, marbles, and balled dynamite. Remembering the effect of the pellets on the coven, she swept them up in her hands and wasted no time crying over Poe.

Sister Sorrow wanted to examine the magic of the man's pellets later on, when she had the time and relaxation period to do so.

Unfortunately, she stumbled on a hook rug in the darkness of one of her multiple rooms and in flinging out her hands to check her fall, dropped all of the curious pellets.

One of them was an explosive.

Which could only be triggered off by a pressure of at least a hundred pounds of weight. That, or the use of a blasting cap.

Catharine Copely weighed one hundred and twenty beautifully arranged pounds. She fell directly across the errant, rolling marble of a pellet. Her full weight bore down on the explosive.

The last thing she ever heard was the keening cry of a siren bansheeing in the night. Like a wail over a graveyard.

It might have been Lucifer screaming.

It wasn't.

It was more probably God laughing.

Whatever deity was represented to Sister Sorrow, the sound was as nothing compared to the tremendous, cataclysmic explosion and consuming fire which turned the penthouse floor of the eyrie on Central Park South into a raging shambles. An inferno of blazing debris.

Columbus Circle had never had a holocaust its equal.

None of the coven got out alive, either.

Later, Philip St George was to regret but one thing.

The death of Mary Vernon. Another runaway heiress.

He would wish a thousand times over that he had taken the girl with him. But he had left her for the police to find. To save and restore to a normal life, whatever her life had been.

EXIT, WITCHCRAFT

A full twenty-four hours later, several related events occurred which had everything to do with the sensational explosion and fire on Central Park South. Miraculously, only the penthouse complex had been destroyed, and there were no casualties other than the assorted charred bodies found on the blasted premises. Identification and the cause of the mysterious holocaust had not yet been determined. The fire and police departments were working overtime on that. It was the greatest calamity, world-wide, that day.

Mr Charles Carmody read the story in his morning *Times*, nearly gagged on his perfecto, canceled all his business appointments for that day, and booked passage on the fastest, most available jet to far-off Hawaii. Carmody, of course, was not a wrigging green garden snake. Merely another of Catharine Copely's willing dupes and love slaves, he had gone along with her deception in her terraced garden, pleased to be taken so far into her confidence. With trick lighting, special effects, and visual aids he had assisted Sister Sorrow in another of her periodic attempts to keep the coven in line, with demonstrations of wizardry and sorcery. Quite effectively, too. The sister had been very pleased; The coven had been awed speechless.

Agents Fling and Lenning, along with the various heads of Homicide, were sorry to have arrived too late to prevent the holocaust. But when the ashes were sifted they did find enough items to securely lock the woman called Catharine Copely with a Devil cult. The ax was found, the paraphernalia, the twisted, fused crucifixes and sundry artifacts. The police lab was not at a loss this time. No matter how many times the ax had been cleaned by the murderer or murderers, blood will tell. Blood which matched two of the five victims of slaughter.

The crank caller, the mystery man, remained a mystery.

Fling and Lenning were sorry about that. Whoever the character was, he deserved a medal. And a bounty fee, perhaps. Or a screen test!

Peter Bond did not go back to the Hotel Americana to check out. To all practical intents and purposes, the Bond characterization was now useless. But he had served his purpose – and well. Too well, perhaps.

So Philip St George appeared at the offices of Kite, Dorn, and Schindler, faultlessly attired, handsome as ever save for an odd bruise on his bronzed forehead. Miss Walters quickly ushered him into the inner office, which was when Sidney Kite raised the roof. The walls reverberated with his ire. Kite read newspapers too and though he had no more facts than the next man, he was convinced that his number-one client had been somewhere in the middle again. Especially when Philip indicated that he would have no need for the mountainous pile of homework acquired for him by Miss Irene Walters and Mr Sidney Kite. Talk about overhead!

The lawyer gave up, however. As he usually did.

He was too happy to see his client alive and well and living in New York. St George Senior would have been proud, too.

'Phil, Phil.'

'Sidney.'

'We work our horns off, you disappear, the town's upside down with this cult being discovered over there on Central Park South, and you walk in and say nothing. Jeezis.'

Philip St George smiled from the depths of the client's chair. His piercing eyes were gentle this time. For his dearest friend.

'It's over, Sidney. And come to the right end. What matters who did what? Those Satan gamesters are out of business. Down at Chic, Incorporated all the clerks and salesgirls are shaking their heads and saying, "*Not Miss Copely! Not her!*" And that's the way it will always be. People hiding behind false fronts and false faces.'

Kite glared at that but didn't refer to it at all. Rather, he had a question of his own. 'What about Carlson? The cyanide? You still on the hook for that one?'

'No. The police will receive a letter in today's afternoon mail. They won't have to guess anymore. I suddenly realized what happened. How it was done. Norma was very nervous, drinking her boilermaker. The rye and the beer. I should have noticed she was chewing a stick of gum. I didn't. In any case, the cyanide had to be in that. The gum wasn't found because she dropped it on the floor when she died. It fell out of her mouth. And I stepped on it, rushing to her. I walked out of that restaurant with a wad of poison gum clinging to the sole of my left shoe. Crazy world, isn't it?'

Sidney Kite blinked. 'When did you find the gum?'

'Last night when I – changed my clothes again – pair of Italian shoes I hadn't worn since the Wallingford.'

'I see. You sent the police a letter. Anonymous letter, of course?'

'Of course. The wad of gum is inside. There will be traces.'

Kite stared at the remarkable young man across from him. The December day was foggy, overcast. The snow was gone but the weather was still lousy. And Philip St George looked like someone just back from a Bermuda cruise. He exuded attractiveness, wealth, and well-being.

Not cockamamie undercover crusades, high adventure and death.

'What am I going to do with you?' Mr Kite almost whimpered.

'What you always do. And so well. Be my mentor, friend, and advisor. But friend above all. I don't know what I'd do without you, Sidney. I mean that. From the bottom of my estate.'

Both men laughed and the harmony of the sound filled the office.

'Phil – it was you again, wasn't it? Tell me the truth now.'

'And shame the Devil?' Philip St George smiled very grimly. 'Yes, Sidney. It was I. In all my cockamamie glory. And I'll do it again and again when I have to. It's as basic and elemental as that.'

'Yeah – well, it will kill you someday. You schnook.'

'Perhaps. But what better work is there for a man like me?'

Sidney Kite had no answer for that, either. Not even a sound Jewish proverb. Sighing, giving up, he reached into the bottom drawer of his desk for a good bottle of Scotch. Ulcer or no, this was worth a drink. A good long one, by way of celebration. And reward.

In the outer office, at her desk, Miss Irene Walters was thinking about Philip St George, in spite of her best efforts not to.

Damn the man. Still Prince Charming with all the stops out.

She had never seen him looking so well.

Small wonder.

Stamping out the Devil and his works was positively the best medicine for the man who was becoming a living legend in his own time. A man spoken of and referred to in all the official police files of the world as – the *Satan Sleuth*.

Peter Bond was dead.

Philip St George was alive.

Long live the *Satan Sleuth*.

MEWS BESTSELLERS

NEL P.O. BOX 11, FALMOUTH, TR10 9EN, CORNWALL.

For U.K.: Customers should include to cover postage, 18p for the first book plus 8p per copy for each additional book ordered up to a maximum charge of 66p.

For B.F.P.O. and Eire: Customers should include to cover postage, 18p for the first book plus 8p per copy for the next 6 and thereafter 3p per book.

For Overseas: Customers should include to cover postage, 20p for the first book plus 10p per copy for each additional book.

Name ....

Address ....

....

....

Title ....

Whilst every effort is made to maintain prices, new editions or printings may carry an increased price and the actual price of the edition supplied will apply.